Climate Resilient Innovative Livelihoods in Indian Sundarban Delta

Abhijit Mitra · Sufia Zaman · Prosenjit Pramanick

Climate Resilient Innovative Livelihoods in Indian Sundarban Delta

Scopes and Challenges

Springer

Abhijit Mitra
Department of Marine Science
University of Calcutta
Kolkata, West Bengal, India

Sufia Zaman
Department of Oceanography
Techno India University
Kolkata, West Bengal, India

Prosenjit Pramanick
Department of Oceanography
Techno India University
Kolkata, West Bengal, India

ISBN 978-3-031-42632-2 ISBN 978-3-031-42633-9 (eBook)
https://doi.org/10.1007/978-3-031-42633-9

This Springer imprint is published by the registered company Springer Nature Switzerland AG
The registered company address is: Gewerbestrasse 11, 6330 Cham, Switzerland

Paper in this product is recyclable.

Acknowledgements

The authors, from the innermost core of mind and heart, acknowledge the following personalities for their direct and indirect contributions.

- **Late Professor Amalesh Choudhury**: All the authors gained in-depth knowledge on mangrove ecology from Late Professor Choudhury not only in the capacity of students, but also as sincere disciples. He taught the authors to work in the rigorous environment of Sundarbans with sincerity, dedication, and tireless effort to reveal the ground reality of the ecosystem.
- **Dr. Goutam Roychowdhury**: The authors greatly acknowledge the infrastructural facilities offered by Techno India University, West Bengal, at Salt Lake Campus while preparing the manuscript. The authors also received inspiration for touching the sky from Sri Roychowdhury. The knowledge of statistical software used to reflect the spatio-temporal variation of abiotic parameters in explaining the biotic community structure in the mangrove ecosystem was offered under the guidance of Sri Roychowdhury. He also inspired the authors to apply AI in forecasting the trend of environmental variables in the framework of the Sundarban ecosystem.

Dr. Abhijit Mitra expresses his gratefulness to his wife Shampa, daughter Ankita, and late mother Manjulika whose inspirations and encouragements act as booster to complete the manuscript. The statement of Late Dhanesh Chandra Mitra, father of Dr. Abhijit Mitra, to create a strong footprint in life still inspires the author with heavenly energy. The sudden demise of the mother of Dr. Mitra during the COVID phase retarded his pace significantly, but with the active support of his well-wisher and Chancellor of Techno India University, West Bengal, Dr. Goutam Roy Chowdhury, the author, could finally complete the manuscript.

Dr. Sufia Zaman expresses her deepest gratitude to her mother Mrs. Ayesha Zaman and her father, Mr. Salim-uz-Zaman, who gave her immense moral support. Dr. Zaman also acknowledges the support of her beloved husband Dr. Sahid Imam Mallick and her baby girl Shanza Mallick. Dr. Zaman wishes to accord her deep sense of gratitude to her family members including her uncle (Mr. Pradip Kumar

Mitra) and aunt (Late Mrs. Kanika Mitra), younger sister (Ms. Sharmilee Zaman), brother-in-law Kazi Dr. Sazzad Manir, her nephew baby Diyan Kazi, her in-laws (Mrs. Zobeda Khatun and Lt. Dr. G. R. Mallick), and her beloved grandmother (Late Mrs. Shibani Dhar) for their encouragement and inspiration throughout the strenuous period of manuscript preparation.

Dr. Prosenjit Pramanick expresses his gratefulness to his father Mr. Amar Sankar Pramanick, mother Mrs. Rekha Pramanick, and aunt Ms. Benodini Samui, who gave him unconditional love, inspiration, support, and enormous encouragement. Dr. Pramanick also acknowledges his sister Mrs. Monalisha Ghosh, brother-in-law Mr. Nilkanta Ghosh, and nephew Mr. Nilax Ghosh for their inspiration to complete the manuscript.

All the authors of this textbook depended on the contributions of several people, mostly those researchers whose sincere labor in the field and laboratory provided the raw materials out of which the chapters have been synthesized. These people are Dr. Tanmay Ray Chaudhuri, IPS, Dr. Mourani Sinha, Dr. Subhra Bikash Bhattacharyya, Dr. Kakoli Bannerjee, Dr. Amitava Aich, Dr. Harekrishna Jana, Dr. Rajrupa Ghosh, Dr. Shankhadeep Chakraborty, Dr. Nabonita Pal, Dr. Arpita Saha, Dr. Sana Ahmed, and Mr. Pavel Biswas.

About This Book

Mangrove forests provide a wide range of provisioning services like timber, fodder, honey, wax, tannin, fish, etc., and a wide range of environmental services like erosion control, protection from natural disasters, bioremediation, biodiversity conservation, carbon sequestration, etc. The ecosystem contains a much larger quantum of stored carbon compared to many terrestrial forests and is also known for its high rate of primary production. The role of mangroves to resist climate-induced sea level rise, salinification, temperature rise, and acidification have been studied in detail, but in context to the Indian Sundarbans delta, their effects are not well defined with data sets.

The livelihoods associated with the mangrove ecosystem at the apex of the Bay of Bengal are also poorly understood. The mangrove forests in deltaic Sundarbans are areas of complexity, opportunity, and conflict (particularly man-animal conflict). They are rich in natural resources, but also fragile in nature due to the impact of climate change that has made the ecosystem highly vulnerable. The deltaic ecosystem is the homeland of about 4.5 million Indian population and the livelihoods of this large percentage of population are mostly oriented around the mangrove forests. Agriculture, pisciculture, aquaculture (preferably shrimp culture), animal husbandry, deep-sea fishing, and prawn seed collection from the estuaries are the traditional livelihoods of the people of Indian Sundarbans. However, many of these livelihoods are detrimental to the environment, particularly in terms of biodiversity loss. Many other dark chapters are also associated with the present mode of livelihoods in the deltaic complex like loss of stored carbon due to the cutting of mangrove flora for timber, fuel, fodder, etc., deterioration of the estuarine water quality due to the proliferation of unplanned shrimp farming at the cost of mangroves, mass destruction of fish juveniles (ichthyoplankton) during the wild collection of prawn seeds, emission of carbon dioxide from the brick kilns and tourist boats, etc. All these activities have accelerated the pulse of climate change in the Indian Sundarban deltaic complex. The island dwellers are facing challenges in their livelihood domains as paddy cultivation, pisciculture, and animal husbandry cannot withstand the intrusion of saline water in the island villages that frequently occurs during natural disasters like super cyclones, tidal surges, etc. To overcome these challenges, it is the need of the hour to bring reformation in the sector of livelihoods. Halophyte farming, polyculture, fish feed preparation from mangrove associate species, and mangrove-based snacks preparation for human consumption are some innovative steps to ensure food security as well as economic security for the people of this fragile ecosystem at the apex of the Bay of Bengal. The market liberalization phase that has spread wings in India can open up new scopes and opportunities for such carbon-neutral mangrove-centric livelihoods. Although many common livelihoods like pisciculture, shrimp farming, and apiculture have been discussed in the framework of Indian Sundarbans, a lot of discrepancies were observed between nascent/practicing entrepreneurs and researchers and entrepreneurs are in line with general criticism that research and practice operate independently of each other. In most of the fundamental resource-based researches, business models are often overlooked, which have been covered for all categories of innovative livelihoods with realistic cost-benefit ratio.

The climate resilient innovative livelihoods addressed in this book have been critically studied through SWOT analysis and emphasized with case studies generated from ground zero observations. The book has its own uniqueness for two reasons. First, mangrove-centric alternative livelihoods with their sustainability have been highlighted and second, the carbon footprint associated with each of the livelihoods has been quantified which can serve as a potential approach to retard the impact of climate change in the region and align with the SDG 14 objectives.

The book critically analyzes several innovative climate resilient alternative livelihoods relevant to the Indian Sundarbans delta through an in-depth SWOT analysis.

We have tried to the best of our ability to make the concepts of the book comprehensive, up to date, and backed up with field level real-time data so that the book may serve as a ready source of information to the readers who want to pursue their careers in Climate Science, Coastal Zone Management, Blue Economy, Environmental Science, Ecology, Oceanography, Marine Science, Fishery, Economics, Geography, and related subjects.

Contents

About the Authors

Dr. Abhijit Mitra Associate Professor and former Head, Department of Marine Science, University of Calcutta (INDIA) has been active in the sphere of Oceanography since 1985. He obtained his Ph.D. as NET qualified scholar in 1994. Since then, he joined Calcutta Port Trust and WWF (World Wide Fund for Nature-India), in various capacities to carry out research programs on environmental science, biodiversity conservation, climate change, and carbon sequestration. Presently, Dr. Mitra is serving as the Director, Research of Techno India University, West Bengal (Hony.). He has to his credit about 685 scientific publications in various National and International journals, and 60 books of postgraduate standards. Dr. Mitra is presently a member of several committees like PACON International, IUCN, SIOS, Mangrove Society of India, The West Bengal National University of Juridical Sciences, etc., and has successfully completed about 17 projects and 19 consultancies on biodiversity loss in the fishery sector, coastal pollution, aquaculture, alternative livelihood, climate change, and carbon sequestration. Dr. Mitra also visited as a faculty member and was invited as a speaker at several Universities in Singapore, Kenya, Oman, and USA. In 2008, Dr. Mitra was invited as a visiting fellow at University of Massachusetts at Dartmouth, USA, to deliver a series of lectures on Climate Change. Dr. Mitra has also successfully guided 54 Ph.D. students. Presently, his research areas include environmental science, mangrove ecology, sustainable aquaculture, alternative livelihood, climate science, carbon sequestration, and blue economy.

Dr. Sufia Zaman presently serving as Head, Department of Oceanography at Techno India University, West Bengal, started her career in the field of Marine Science since 2001. She worked in the rigorous mangrove ecosystem of Indian Sundarbans and has a wide range of experience in exploring the floral and faunal diversities of Sundarbans. She has published six books on carbon sequestration, 310 scientific papers, and contributed chapters in several books on biodiversity, environmental science, aquaculture, and livelihood development. Dr. Zaman is presently a member of the Fisheries Society of India. She is also running projects on carbon sequestration by mangroves of Indian Sundarbans. She is the recipient of DST Women Scientist and

Jawaharlal Memorial Doctoral fellowship awards. Her areas of research include aquaculture, fish nutrition, phytoplankton diversity, climate change mangrove ecology, and alternative livelihood. Dr. Zaman is also the first researcher in the maritime state of West Bengal (India), who initiated trial experiments on iron fertilization and subsequent enhancement of primary (phytoplankton) and secondary (fish) productions in the brackish water ponds of Indian Sundarbans with the financial assistance of Department of Science and Technology, Government of India. Dr. Zaman is also providing consultancies on green technology and emission control to several industries, NGOs, and corporate sectors.

Dr. Prosenjit Pramanick is presently holding the position of Post-doctoral Researcher, Department of Oceanography, Techno India University, West Bengal. He had passed Bachelor of Science in Biochemistry in 2010 from the University of Calcutta, West Bengal, and Master of Science in Biochemistry (in 2012) from Vidyasagar University, West Bengal, and then obtained his Ph.D. degree in 2017 from Techno India University, West Bengal. He has to his credit about 113 scientific publications, 23 book chapters published by Springer International, Emerald Group, CSIR-NISCAIR, two Books, and 18 publications in conference proceedings in the sphere of Food technology, Aquaculture, Agribiotechnology, Alternative livelihood, and Environmental Science. He has participated in various National and International conferences, seminars, workshops, and webinars. He also served as an Associate Editor in an Inter-continental Webinar Proceedings on "Ecosystem Services and United Nations Sustainable Development Goals (UNSDG)" entitled "Natural Resources and their Ecosystem Services". He has expertise in water quality assessment (in situ and ex situ), food product preparation from mangrove floral species and carbon sequestration. He has served in several consultancy services on the carbon sequestration and Biodiversity assessment conducted by Konnagar Municipality, Government of West Bengal, Kingston Educational Institute, Forest Department, Government of West Bengal, Tata Steel Ltd. Kalinganagar, Odisha, India, Tata Steel Ltd. Noamundi Mines, Jharkhand, India, and several coal-based power sectors. He also rendered his technological expertise as Research Associate in a project on innovative fish feed conducted by the Directorate of Fishery, Government of West Bengal.

Chapter 1
Indian Sundarban Delta

Contents

1.1 Physiography

Sundarban is a mangrove dominated delta complex that spreads both in India and Bangladesh. The delta lies at the apex of Bay of Bengal and is formed through sediment deposition by three major rivers namely the Ganges, the Brahmaputra, and the Meghna (Fig. 1.1). This delta is active in nature with an area of about 10,000 sq. km of which approximately 6000 sq. km falls within Bangladesh and about 4000 sq. km lies within the Indian territory. The present chapter focuses on the mangrove dominated Sundarbans in the Indian part.

The delta complex at the confluence of the Hooghly-Matla estuary and the Bay of Bengal sustains the famous mangrove ecosystem of Indian Sundarbans with a Biosphere Reserve area of 9630 sq km. The region sustains a wide range of floral and faunal diversities (Tables 1.1 and 1.2), and is noted for the co-existence of Royal Bengal Tiger (*Panthera tigris tigris*) and halophytes. We observed fresh pug marks of tiger during our ground-zero data collection period from the region (Fig. 1.2).

A. Mitra et al., *Climate Resilient Innovative Livelihoods in Indian Sundarban Delta*,
https://doi.org/10.1007/978-3-031-42633-9_1

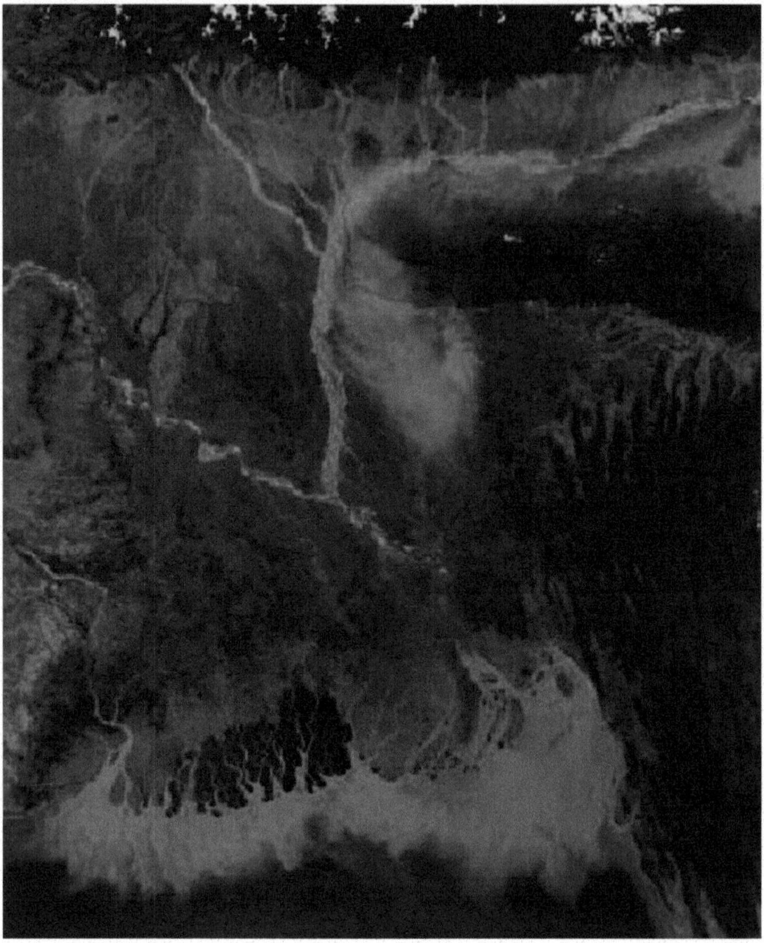

Fig. 1.1 Map showing the location of Sundarbans—spread in both India and Bangladesh

Table 1.1 Floral spectra of Indian Sundarbans

Sl. no	Flora	No. of genus	No. of species	References
1	Bacteria	37	–	Sengupta and Mitra (2020)
2	Fungi	5	–	Danda et al. (2017)
3	Virus	1	–	Danda et al. (2017)
4	Algae	–	270	Danda et al. (2017)
5	Diatoms	–	71	Mitra (2013)
6	Lichens	–	167	Danda et al. (2017)
7	Mangrove and mangrove associate flora	–	178	Mitra (2020)

Table 1.2 Faunal spectra of Indian Sundarbans

Sl. no	Fauna	No. of species	References
1	Protozoa	67	Danda et al. (2017)
2	Copepods	52	Danda et al. (2017)
3	Mollusca	177	Danda et al. (2017)
4	Polychaetes	57	Danda et al. (2017)
5	Xiphosurans	2	Mitra (2020)
6	Crustacea	329	Danda et al. (2017)
7	Spiders	114	Danda et al. (2017)
8	Mites	121	Danda et al. (2017)
9	Insects	497	Danda et al. (2017)
10	Fish	285	Mitra (2020)
11	Amphibia	11	Danda et al. (2017)
12	Reptiles	71	Danda et al. (2017)
13	Aves	250	Mitra (2020)
14	Mammals	47	Danda et al. (2017)

Fig. 1.2 Fresh pug marks of tigers on the intertidal mudflats in Indian Sundarbans

The Indian Sundarban occupies 38% share of the total Sundarbans and the remaining 62% lies within the Bangladesh territory that lies at the east of the Indian part of Sundarbans. The region is noted for the presence of seven major estuaries (Mitra 2013), but more important is the significant variation of water quality during two tidal phases (HT and LT). Tables 1.3 and 1.4 reflect the variations of hydrological parameters at twelve selected sampling stations (Table 1.5 and Fig. 1.3) distributed in and around Indian Sundarbans.

Table 1.3 Physico-chemical characteristics of surface water at High Tide (HT) condition during May, 2022

Station no	Water temperature (°C)	Salinity (psu)	pH	Alkalinity (mg/l)	DO (mg/l)
1	36.8	2.93	7.80	154	5.34
2	36.8	4.74	7.61	164	5.05
3	36.9	6.45	7.65	168	6.65
4	36.9	12.95	7.96	193	6.18
5	36.7	12.04	8.05	166	4.42
6	36.8	13.6	7.95	176	4.54
7	36.9	15.19	8.06	206	4.44
8	36.8	15.5	8.05	190	4.38
9	36.9	16.83	8.05	216	4.16
10	36.7	19.14	8.09	255	4.32
11	36.9	20.36	8.15	266	4.66
12	36.7	23.85	8.16	284	4.65

Table 1.4 Physico-chemical characteristics of surface water at Low Tide (LT) condition during May, 2022

Station no	Water temperature (°C)	Salinity (psu)	pH	Alkalinity (mg/l)	DO (mg/l)
1	36.7	2.08	7.55	137	5.63
2	36.7	3.45	7.5	152	5.35
3	36.6	4.78	7.51	151	6.42
4	36.7	11.76	7.8	177	6.61
5	36.5	10.84	7.85	150	4.53
6	36.8	12.3	7.74	160	4.69
7	36.8	13.66	7.95	189	4.82
8	36.6	13.82	7.95	175	4.64
9	36.8	15.3	7.95	198	4.69
10	36.7	16.79	7.94	237	4.49
11	36.8	18.29	8.01	253	4.75
12	36.8	21.92	8.13	265	4.82

Table 1.5 Local name and coordinates of sampling stations

Sl. no	Sampling station	Coordinates
1	Raichak (Stn. 1)	22°12′12.00″N and 88°07′42.09″E
2	Diamond Harbour (Stn. 2)	22°11′04.02″N and 88°10′50.52″E
3	Kulpi (Stn. 3)	22°36′28.86″N and 88°23′28.32″E
4	Balari (Stn. 4)	22°07′02.16″N and 88°01′35.34″E
5	Haldi River mouth (Stn. 5)	22°00′26.07″N and 88°03′29.64″E
6	Nayachar (Stn. 6)	22°00′30.42″N and 88°03′32.52″E
7	Khejuri Reserve Forest (Stn. 7)	21°54′51.66″N and 88°00′56.52″E
8	Ghoramara Island (Stn. 8)	21°56′15.24″N and 88°07′33.06″E
9	Harwood point (Stn. 9)	21°56′15.24″N and 88°07′33.60″E
10	Harinbari (Stn. 10)	21°46′54.12″N and 88°04′02.64″E
11	Chemaguri (Stn. 11)	21°39′49.32″N and 88°09′11.88″E
12	Sagar South (Stn. 12)	21°39′04.68″N and 88°01′47.28″E

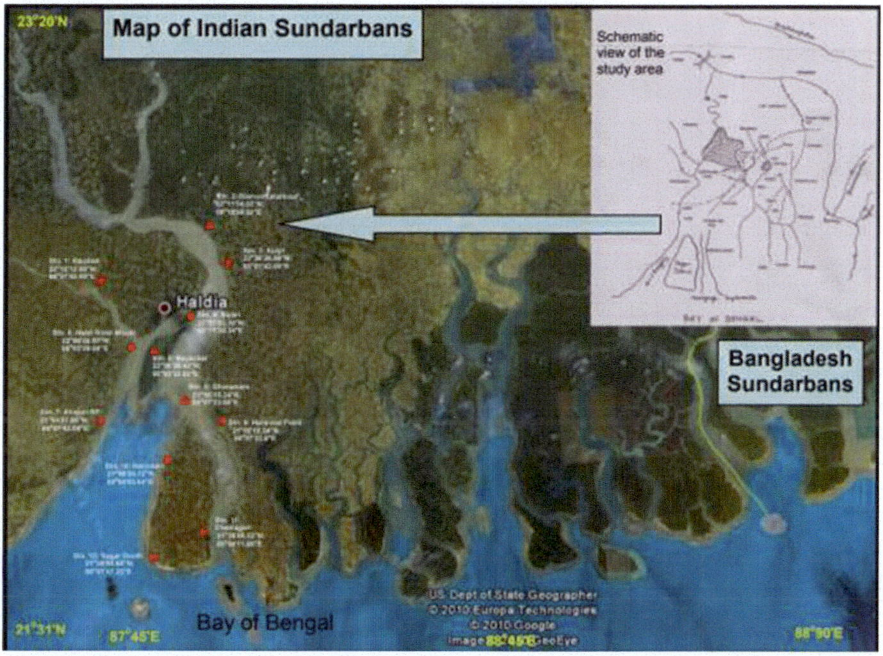

Fig. 1.3 Map showing the lower stretch of Gangetic delta and location of sampling stations

The soil of Indian Sundarbans is saline in nature with a mixed texture of sand, silt, and clay. A total of ten stations distributed in the western (Stn. 1–Stn. 5) and central (Stn. 6–Stn. 10) sectors of Indian Sundarbans (Table 1.6 and Fig. 1.4) were studied by the present authors during January 2021, 2022 and 2023 (Tables 1.7, 1.8, 1.9 and Figs. 1.5–1.13).

The three components sand, silt, and clay in varying proportions, determine the texture of soil. Sand particles are the largest of the three measuring from 1/50th of an inch to 1/500th. Silt particles range from 1/500th of an inch to 1/12500th, and clay particles are less than 1/12500th of an inch—so small that cannot be see with an ordinary microscope. Sand and silt particles are chemically stable in the sense

Table 1.6 Sampling stations with salient features

Study site	Longitude & Latitude	Site description
Harinbari (Stn. 1)	88°04′22.88″ E 21°46′53.07″ N	Situated in the western region of Indian Sundarbans almost in the middle of the Sagar Island; receives the water of the Hugli River
Chemaguri (Stn. 2)	88°08′49.01″ E 21°39′42.88″ N	Situated on the south-eastern side of Sagar Island and receives the water of the Mooriganga River
Sagar South (Stn. 3)	88°04′0.51″ E 21°37′49.90″ N	Situated on the south-western part of the Sagar Island at the confluence of the River Hugli and the Bay of Bengal. Anthropogenically stressed zone due to presence of passenger jetties, fishing activities and pilgrimage
Lothian island (Stn. 4)	88°19′8.47″ E 21°39′08.04″ N	Situated east of Bakkhali; a Wildlife sanctuary; faces the River Saptamukhi
Prentice island (Stn. 5)	88°17′3.62″ E 21°42′43.31″ N	Situated north of Lothian Island; receives the water of the Saptamukhi River
Canning (Stn. 6)	88°41′04.43″ E 22°19′03.20″ N	Situated in the central part of the Indian Sundarbans and faces the mighty River Matla, a tide-fed river. Due to presence of fish landing stations, passenger jetties and busy market, the area is anthropogenically stressed
Sajnekhali (Stn. 7)	88°48′15.78″ E 22°06′34.19″ N	A Wildlife Sanctuary and a part of Sundarban Tiger Reserve; adjacent to River Bidhya and Gomor. Tourism pressure is extremely high in this station particularly during postmonsoon
Chotomollakhali (Stn. 8)	88°54′26.71″ E 22°10′40.00″ N	Situated in the upper portion of Central Indian Sundarban adjacent to Jhilla forest; receives the water of Rangabelia and Korankhali Rivers
Satjelia (Stn. 9)	88°52 ′49.51″ E 22°05 ′17.86″ N	Situated adjacent to River Duttar in the upper region of Central Indian Sundarban facing western part of the Jhilla forest
Pakhiralaya (Stn. 10)	88°48′29.00″ E 22°07′07.23″ N	Situated adjacent to River Gomor; opposite to Sajnekhali Wild Life Sanctuary

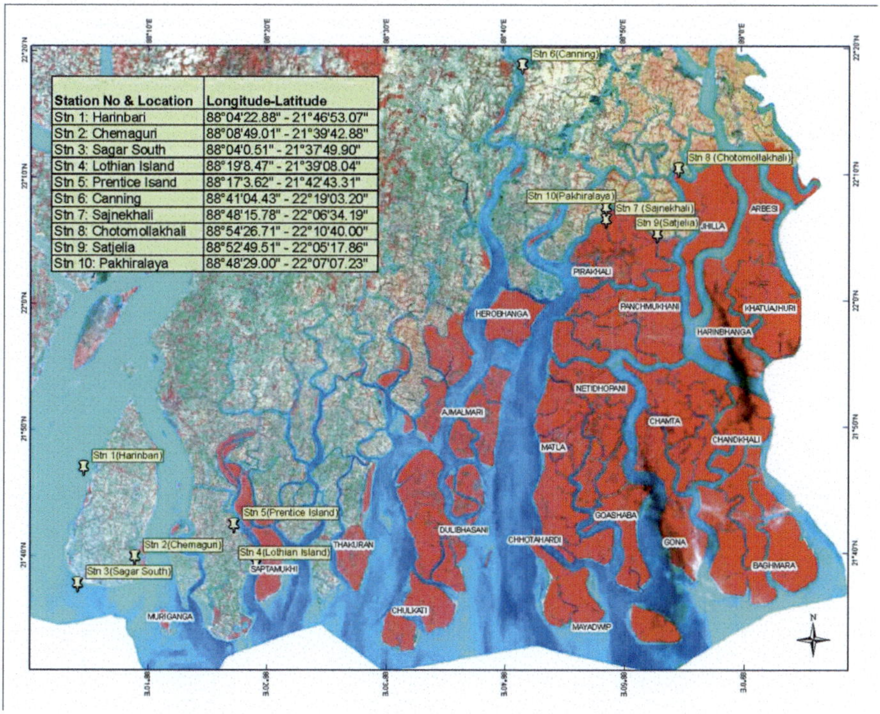

Fig. 1.4 Selected stations in Indian Sundarbans

Table 1.7 Sand, silt and clay percentage in 10 stations of Indian Sundarbans during 2021

Sl No.	Name	Sand (%)	Silt (%)	Clay (%)
1.	Harinbari (Stn. 1)	44.2	30.6	25.2
2.	Chemaguri (Stn. 2)	45.6	30.5	23.9
3.	Sagar South (Stn. 3)	54.7	25.2	20.1
4.	Lothian island (Stn. 4)	42.1	31.1	26.8
5.	Prentice island (Stn. 5)	42.4	30.9	26.7
6.	Canning (Stn. 6)	47.1	27.0	26.9
7.	Sajnekhali (Stn. 7)	41.7	32.0	26.3
8.	Chotomollakhali (Stn. 8)	43.0	32.7	24.3
9.	Satjelia (Stn. 9)	45.3	29.9	24.8
10.	Pakhiralaya (Stn. 10)	43.8	30.0	26.2

that they remain the same chemical composition as the mother rock from which they came. In the present study, it is found that in all the selected stations the order is sand > silt > clay. It is also noted that sand percentage is highest in Sagar South (54.7% in 2021, 55.4% in 2022 and 55.6% in 2023), which may be due to its location at the

Table 1.8 Sand, silt and clay percentage in 10 stations of Indian Sundarbans during 2022

Sl No.	Name	Sand (%)	Silt (%)	Clay (%)
1.	Harinbari (Stn. 1)	45.0	30.1	24.9
2.	Chemaguri (Stn. 2)	45.7	30.0	24.3
3.	Sagar South (Stn. 3)	55.4	24.4	20.2
4.	Lothian island (Stn. 4)	42.8	31.0	26.2
5.	Prentice island (Stn. 5)	42.3	30.8	26.9
6.	Canning (Stn. 6)	46.2	27.4	26.4
7.	Sajnekhali (Stn. 7)	41.8	32.1	26.1
8.	Chotomollakhali (Stn. 8)	43.5	32.5	24.0
9.	Satjelia (Stn. 9)	45.5	29.8	24.7
10.	Pakhiralaya (Stn. 10)	43.7	29.9	26.4

Table 1.9 Sand, silt and clay percentage in 10 stations of Indian Sundarbans during 2023

Sl No.	Name	Sand (%)	Silt (%)	Clay (%)
1.	Harinbari (Stn. 1)	44.5	30.2	25.3
2.	Chemaguri (Stn. 2)	45.3	30.1	24.6
3.	Sagar South (Stn. 3)	55.6	24.1	20.3
4.	Lothian island (Stn. 4)	42.4	31.3	26.3
5.	Prentice island (Stn. 5)	42.6	30.6	26.8
6.	Canning (Stn. 6)	46.0	27.6	26.4
7.	Sajnekhali (Stn. 7)	41.3	32.2	26.5
8.	Chotomollakhali (Stn. 8)	43.4	32.5	24.1
9.	Satjelia (Stn. 9)	45.6	29.9	24.5
10.	Pakhiralaya (Stn. 10)	43.3	30.0	26.7

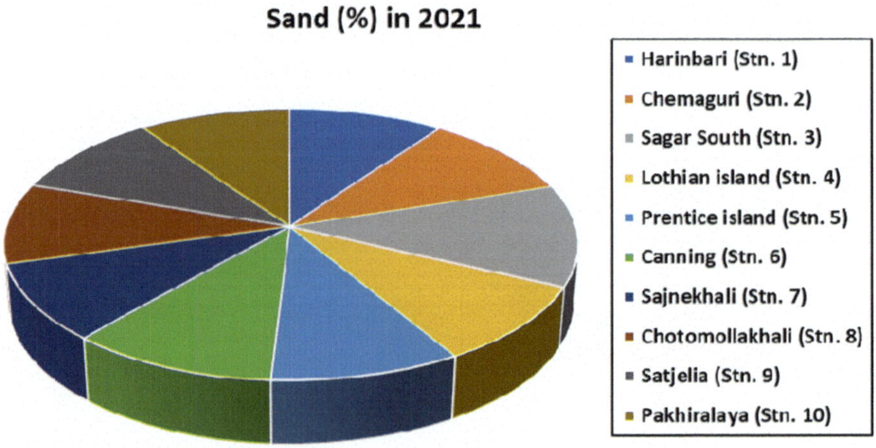

Fig. 1.5 Sand (%) in ten selected stations in Indian Sundarbans during 2021

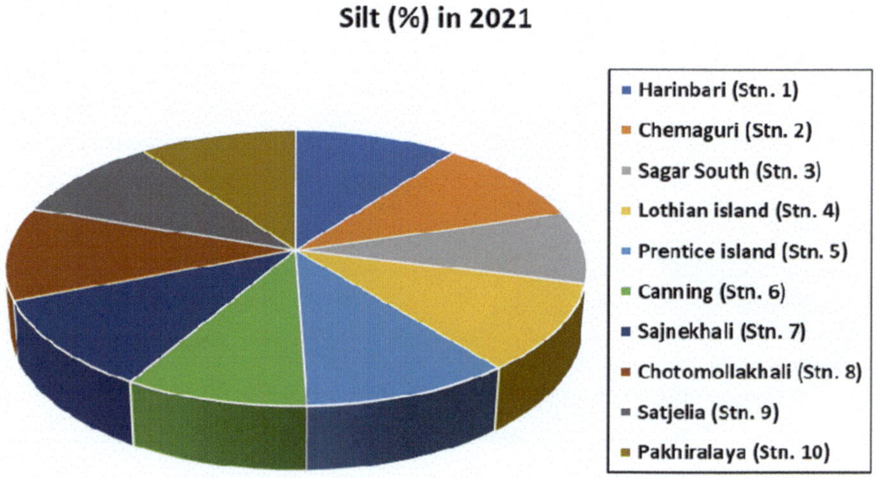

Fig. 1.6 Slit (%) in ten selected stations in Indian Sundarbans during 2021

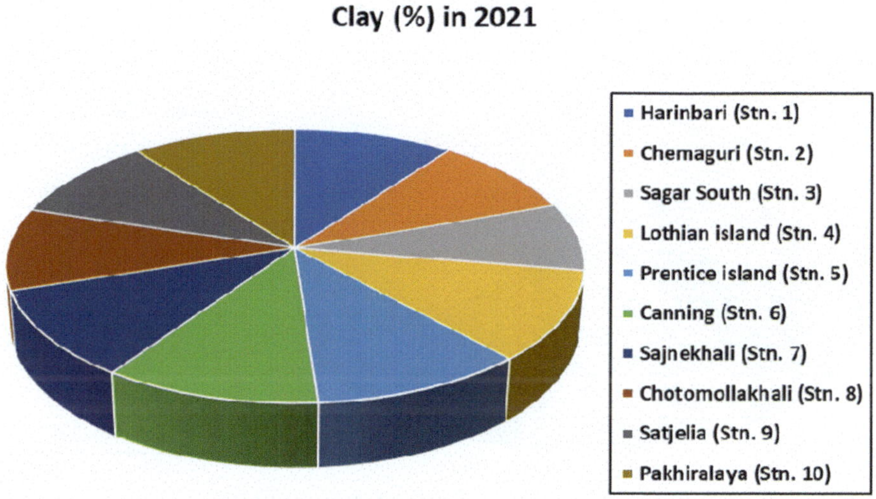

Fig. 1.7 Clay (%) in ten selected stations in Indian Sundarbans during 2021

confluence of Bay of Bengal and the Hooghly River. The high tide from the Bay of Bengal brings lots of sand particles that get deposited in the littoral zone (Fig. 1.14).

The percentage of silt and clay are higher in two stations of western Indian Sundarbans (Lothian Island and Prentice Island), which is the effect of deposition of detritus from the rich mangrove vegetation in these two islands (Fig. 1.15).

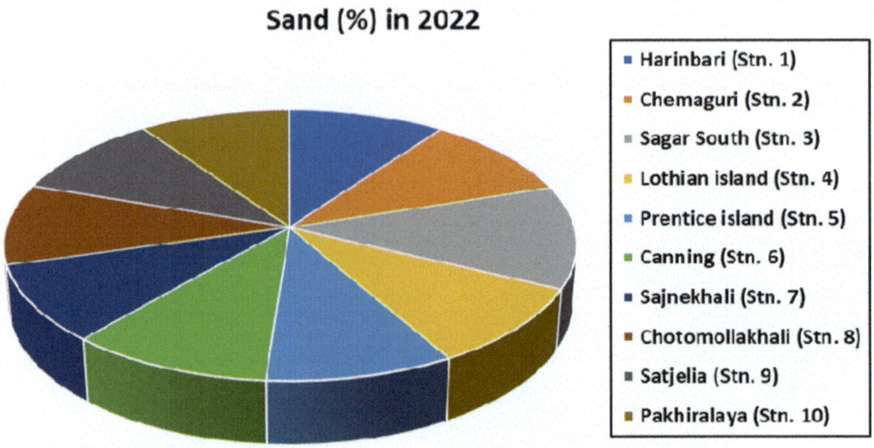

Fig. 1.8 Sand (%) in ten selected stations in Indian Sundarbans during 2022

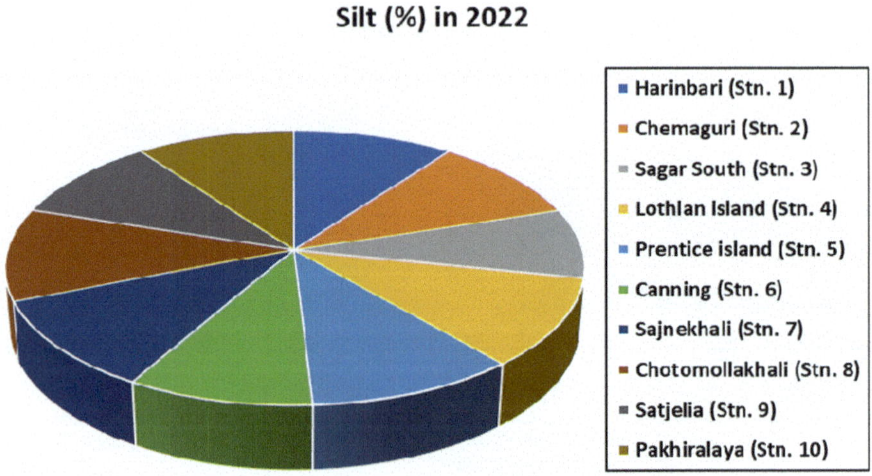

Fig. 1.9 Slit (%) in ten selected stations in Indian Sundarbans during 2022

Siltation also regulates the survival and growth of mangrove species. Anatomical differences are observed between highly and less silted sites. The stomatal area is considerably lower for high siltation site. Considering the ten stations as stated in Tables 1.7, 1.8 and 1.9, it is observed that silt is relatively higher in stations like Harinbari, Chemaguri, Lothian Island, Prentice Island, Sajnekhali and Chotomollakhali compared to Sagar South, Canning, Satjelia and Pakhiralaya (*vide* Tables 1.7, 1.8 and 1.9 for the years 2021, 2022 and 2023 respectively for data on sand, silt, and clay percentage).

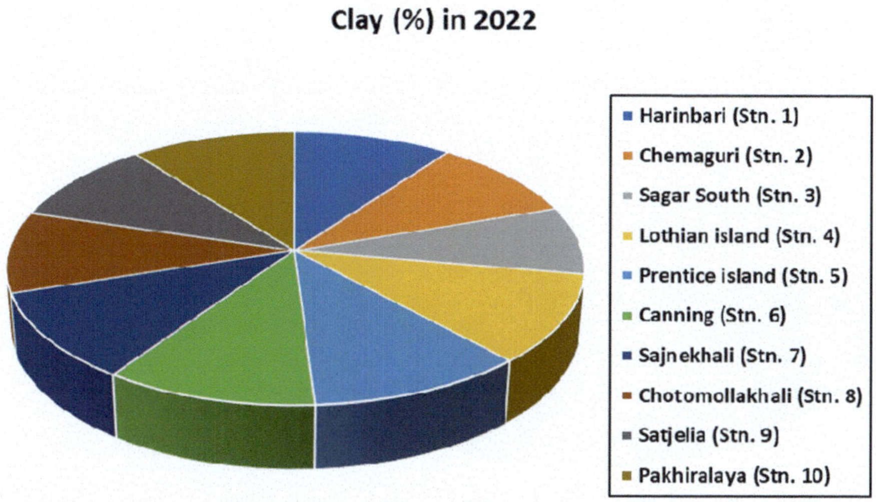

Fig. 1.10 Clay (%) in ten selected stations in Indian Sundarbans during 2022

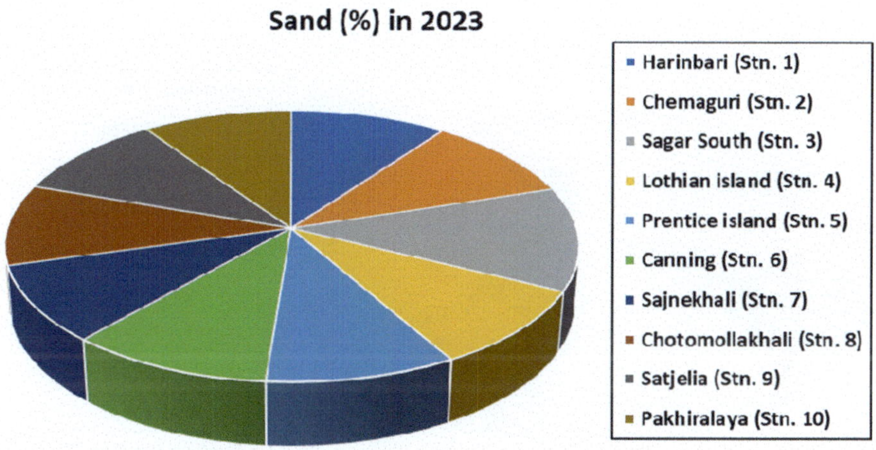

Fig. 1.11 Sand (%) in ten selected stations in Indian Sundarbans during 2023

The halophytes present in the ecosystem are the sources of organic carbon although anthropogenic activities like shrimp farming, discharges from tourism units and fish landing stations generate considerable amount of organic matter in the ambient media.

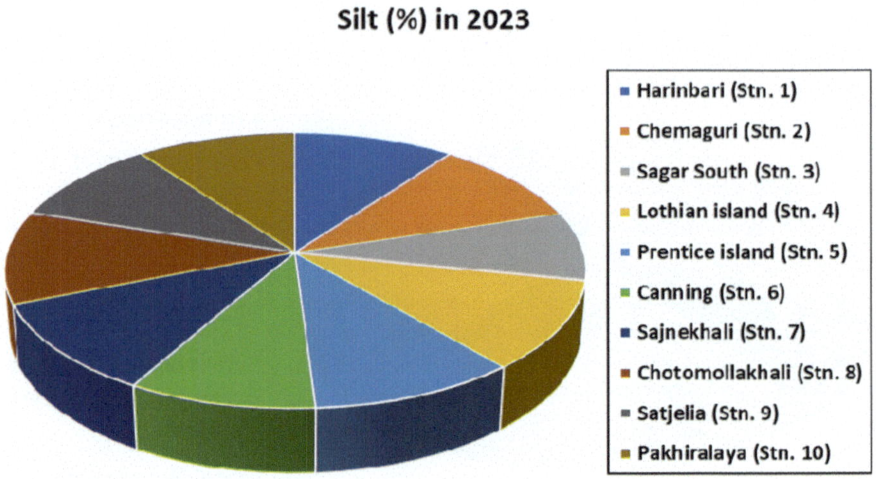

Fig. 1.12 Slit (%) in ten selected stations in Indian Sundarbans during 2023

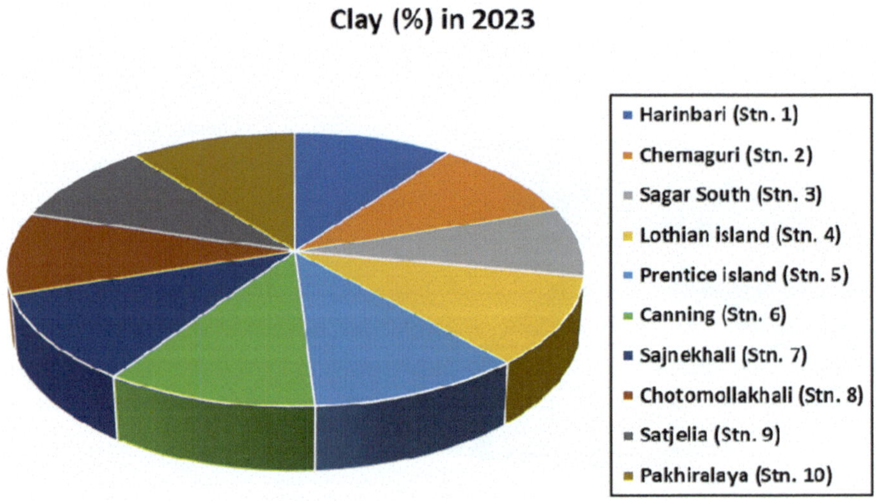

Fig. 1.13 Clay (%) in ten selected stations in Indian Sundarbans during 2023

We carried out a survey during May 2021, 2022 and 2023 in the above 10 stations (Table 1.10) to monitor the organic carbon in the surface soil of the selected stations distributed in the western and central Indian Sundarbans. We observed significant spatial variations, which may be attributed to variation in relative abundance of mangroves and several other anthropogenic activities common in the study area.

Fig. 1.14 High tides fetching sand from the offshore to the intertidal zone

Fig. 1.15 Litters and detritus contributed by the mangrove flora

Table 1.10 SOC (in %) in 10 stations of Indian Sundarbans

Sl. No.	Station Name	2021	2022	2023
1.	Harinbari (Stn. 1)	1.42	1.40	1.49
2.	Chemaguri (Stn. 2)	1.29	1.27	1.34
3.	Sagar South (Stn. 3)	0.80	0.87	0.90
4.	Lothian island (Stn. 4)	1.37	1.35	1.41
5.	Prentice island (Stn. 5)	1.36	1.39	1.38
6.	Canning (Stn. 6)	0.83	0.85	0.87
7.	Sajnekhali (Stn. 7)	1.30	1.32	1.36
8.	Chotomollakhali (Stn. 8)	0.93	0.97	1.01
9.	Satjelia (Stn. 9)	1.13	1.15	1.21
10.	Pakhiralaya (Stn. 10)	1.24	1.29	1.33

Soil Organic Carbon (SOC) ranged from 0.80% to 1.49% during the study period (Table 1.10). It is observed that the SOC level is more in the western Indian Sundarbans compared to the central sector. It is interesting to note that although the station Sagar South is situated in the western Indian Sundarbans, but the value of SOC percentage is very low (0.80%). This may be due to continuous erosion of the inter-tidal mud flats in this station located at the confluence of the River Hooghly and the Bay of Bengal. The high values of SOC in stations like Harinbari (1.42%) and Chemaguri (1.29%) may be due to high anthropogenic influences in these two stations. Again, it is also observed that SOC levels are high in Lothian Island (1.40%) and Prentice Island (1.39%), although these areas are pristine and are devoid of any anthropogenic influences. The relatively higher value of SOC in these two stations may be due to presence of more mangrove diversity with greater biomasses. The lower values of SOC in central Indian Sundarbans may be the effect of stunted growth of trees due to high salinity which has resulted in reduction of mangrove litter production that serves as natural source material of SOC in the mangrove ecosystem (Mitra et al. 2010; Banerjee et al. 2010; Mitra et al. 2011; Mitra 2013; Jana et al. 2014; Mitra et al. 2016; Trivedi et al. 2016; Banerjee et al. 2017; Mitra 2018; Agarwal and Mitra 2018; Mitra 2018; Guha and Mitra 2020; Dhar et al. 2021, Mitra et al. 2022).

Tidal variation is a vital component in deltaic Sundarbans as it not only regulates the halophytic vegetation pattern, but also the livelihood profile of the island dwellers. We carried out observations at 24 stations in Indian Sundarban estuaries (Fig. 1.16) from boats that were anchored for this tide related research programme during 2021. Our observations started at 5 am (Indian Standard Time or IST) on 22nd April 2021 and ended at 5 am on 25th April 2021. The results of these 72 h. observational periods are presented in Table 1.11 and Fig. 1.17.

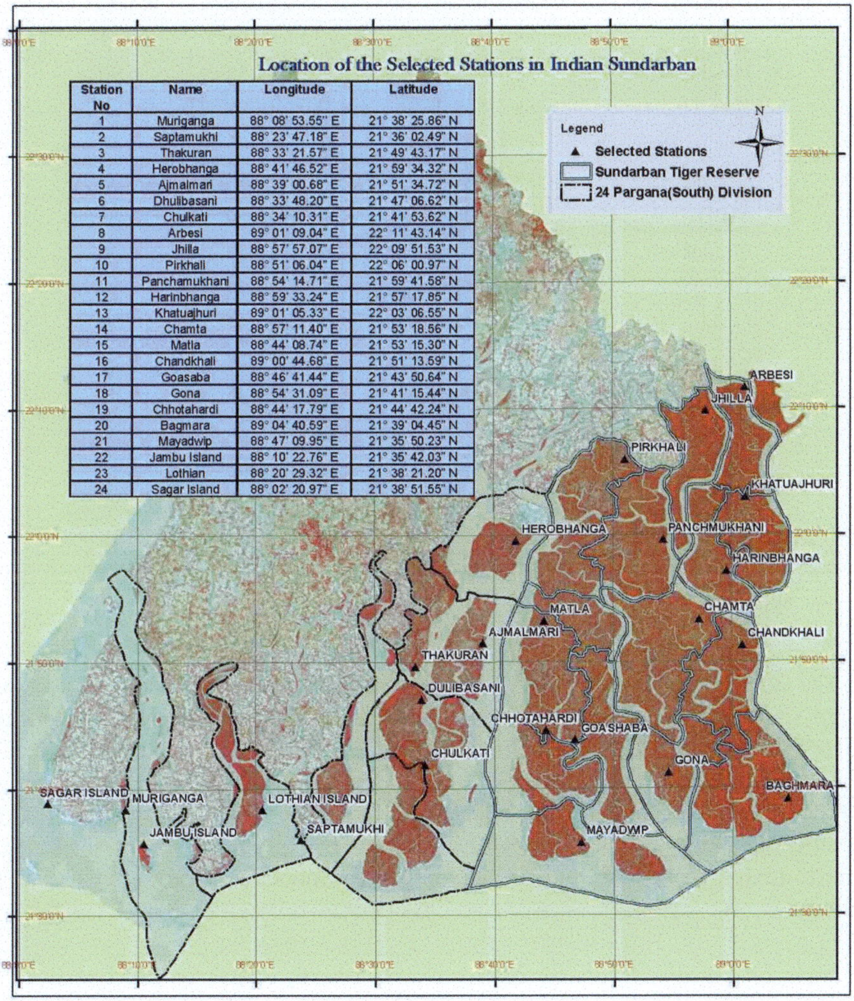

Station No	Name	Longitude	Latitude
1	Muriganga	88° 08' 53.55" E	21° 38' 25.86" N
2	Saptamukhi	88° 23' 47.18" E	21° 36' 02.49" N
3	Thakuran	88° 33' 21.57" E	21° 49' 43.17" N
4	Herobhanga	88° 41' 46.52" E	21° 59' 34.32" N
5	Ajmalmari	88° 39' 00.68" E	21° 51' 34.72" N
6	Dhulibasani	88° 33' 48.20" E	21° 47' 06.62" N
7	Chulkati	88° 34' 10.31" E	21° 41' 53.62" N
8	Arbesi	89° 01' 09.04" E	22° 11' 43.14" N
9	Jhilla	88° 57' 57.07" E	22° 09' 51.53" N
10	Pirkhali	88° 51' 06.04" E	22° 06' 00.97" N
11	Panchamukhani	88° 54' 14.71" E	21° 59' 41.58" N
12	Harinbhanga	88° 59' 33.24" E	21° 57' 17.85" N
13	Khatuajhuri	89° 01' 05.33" E	22° 03' 06.55" N
14	Chamta	88° 57' 11.40" E	21° 53' 18.56" N
15	Matla	88° 44' 08.74" E	21° 53' 15.30" N
16	Chandkhali	89° 00' 44.68" E	21° 51' 13.59" N
17	Goasaba	88° 46' 41.44" E	21° 43' 50.64" N
18	Gona	88° 54' 31.09" E	21° 41' 15.44" N
19	Chhotahardi	88° 44' 17.79" E	21° 44' 42.24" N
20	Bagmara	89° 04' 40.59" E	21° 39' 04.45" N
21	Mayadwip	88° 47' 09.95" E	21° 35' 50.23" N
22	Jambu Island	88° 10' 22.76" E	21° 35' 42.03" N
23	Lothian	88° 20' 29.32" E	21° 38' 21.20" N
24	Sagar Island	88° 02' 20.97" E	21° 38' 51.55" N

Fig. 1.16 Location of sampling stations

Table 1.11 Sampling stations in Indian Sundarban estuarine mudflats

Sl. no	Sampling station	Longitude	Latitude	Tidal range
1	Muriganga	88°08′53.55″E	21°38′25.86″N	5.05
2	Saptamukhi	88°23′47.18″E	21°36′02.49″N	4.58
3	Thakuran	88°33′21.57″E	21°49′43.17″N	4.35
4	Herobhanga	88°41′46.52″E	21°59′34.32″N	4.28
5	Ajmalmari	88°39′00.68″E	21°51′34.72″N	4.54
6	Dhulibasani	88°33′48.20″E	21°47′06.62″N	4.31
7	Chulkathi	88°34′10.31″E	21°41′53.62″N	4.26
8	Arbesi	89°01′09.04″E	22°11′43.14″N	4.48
9	Jhilla	88°57′57.07″E	22°09′51.53″N	4.34
10	Pirkhali	88°51′06.04″E	22°06′00.97″N	4.15
11	Panchmukhani	88°54′14.71″E	21°59′41.58″N	4.90
12	Harinbhanga	88°59′33.24″E	21°57′17.85″N	4.85
13	Khatuajhuri	89°01′05.33″E	22°03′06.55″N	4.47
14	Chamta	88°57′11.40″E	21°53′18.56″N	4.55
15	Matla	88°44′08.74″E	21°53′15.30″N	5.01
16	Chandkhali	89°00′44.68″E	21°51′13.59″N	4.36
17	Goashaba	88°46′41.44″E	21°43′50.64″N	4.52
18	Gona	88°54′31.09″E	21°41′15.44″N	4.63
19	Chhotahardi	88°44′17.79″E	21°44′42.24″N	4.27
20	Baghmara	89°04′40.59″E	21°39′04.45″N	5.04
21	Mayadwip	88°47′09.95″E	21°35′50.23″N	4.96
22	Jambu Island	88°10′22.76″E	21°35′42.03″N	5.12
23	Lothian	88°20′29.32″E	21°38′21.20″N	4.95
24	Sagar Island	88°02′20.97″E	21°38′51.55″N	5.15

* Tidal ranges at any location are defined as the differences between successive high water (HW) and low water (LW)

The average tidal range in this region oscillates around 4.5 m, but during cyclonic depressions that frequently occur in the delta region, the upper range of tide touches up to 9 m due to which the agricultural fields, freshwater ponds/canals get salinized. This causes great loss to the people of the region in terms of their livelihoods as paddy, endemic vegetables of the islands and freshwater fishes (preferably the carp varieties) cannot withstand saline water. On this background, it is the need of the hour to review the possibilities of alternative livelihoods with strong resilience to salinity.

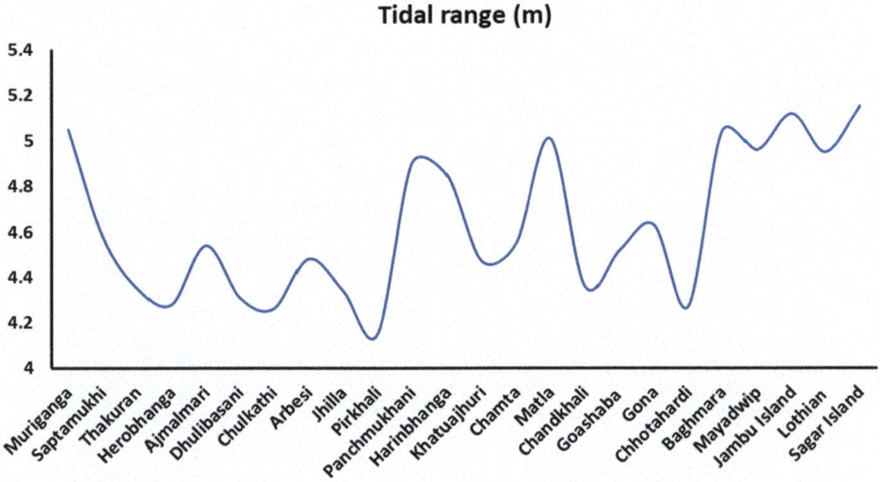

Fig. 1.17 Tidal range observed in 24 selected station of Indian Sundarbans during our survey period

1.2 Demographic Profile

Sundarban deltaic complex is known for its unique diversity of halophytes. The region is bounded by Bangladesh in the east, the Hooghly River (a continuation of the River Ganga) in the west, the Dampier and Hodges line in the north, and the Bay of Bengal in the south. The reclamation history shows that man had started to settle in this delta region since 1770 AD in the early part of colonial British rule. The distribution of population in this region has a relationship with the natural and socio-economic factors, which are unevenly distributed over the surface of the region. This unit attempts to show the distribution pattern of the population. The population distribution encompassing total population, male and female population are not uniform throughout all the 19 blocks in Indian Sundarbans (Figs. 1.18, 1.19 and 1.20).

The census records (https://www.sundarbanaffairswb.in/home/block) show that the ration between the male and female population in Indian Sundarbans is 1000:955. It is to be noted that a large chunk of the population in the region belong to Scheduled Caste (SC) and Scheduled Tribe (ST) as revealed in Figs. 1.21 and 1.22.

The people dwelling in Indian Sundarbans have a very narrow band of livelihoods. There is not much choice except cultivation and fishing. The area specific approach and regional plan for the delta complex is of utmost importance as the region is suffering from multiple problems like man-animal conflict, increased salinity of the water and erosion of embankment.

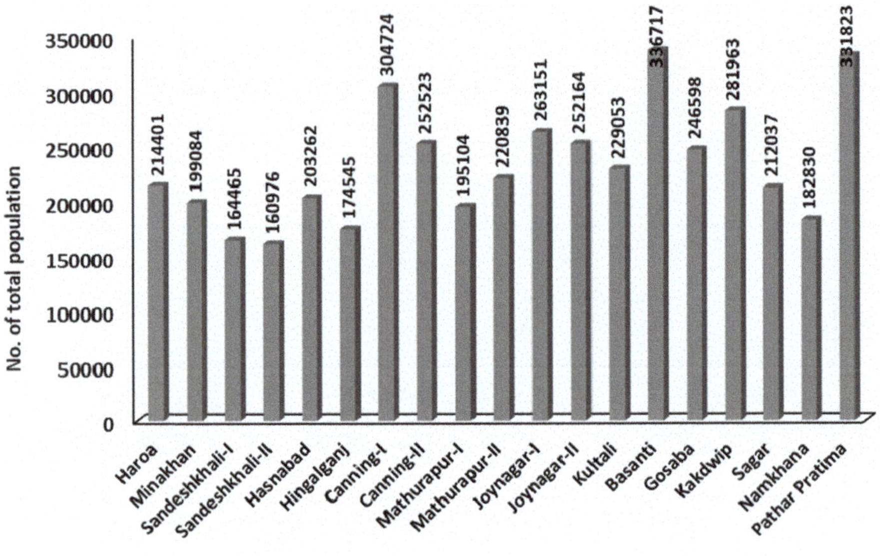

Fig. 1.18 Block-wise total population in Indian Sundarbans

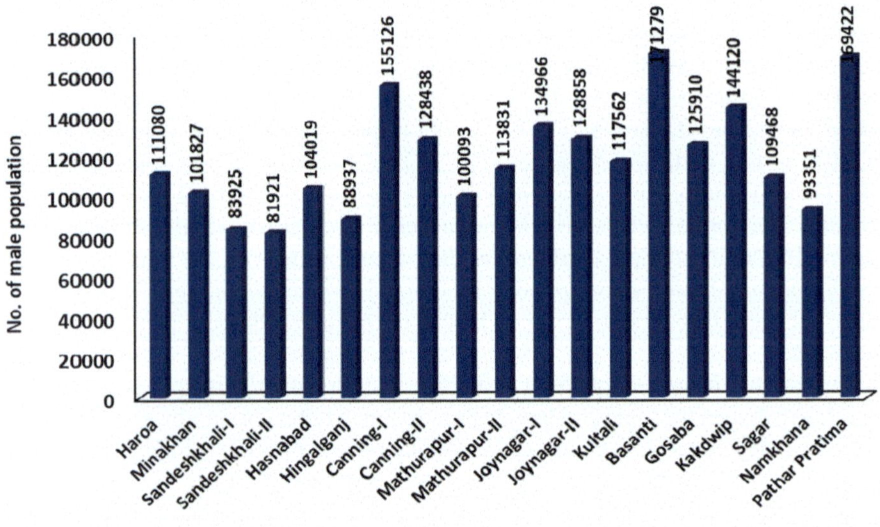

Fig. 1.19 Block-wise male population in Indian Sundarbans

1.3 Climatic Trend

Climate change has posed significant impacts on the marine and estuarine ecosystems. The pulse of climate change is felt even in the pristine parts of Indian Sundarbans where there are no industries or any anthropogenic impacts of considerable

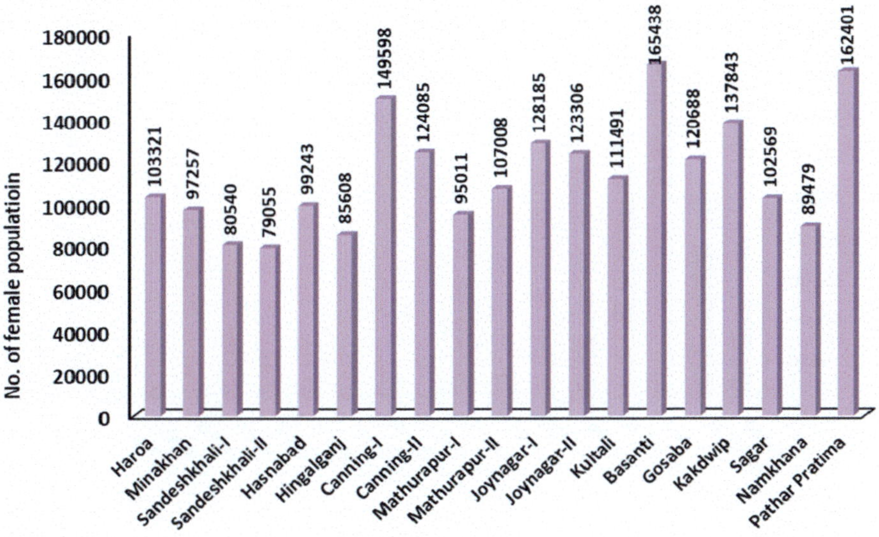

Fig. 1.20 Block-wise female population in Indian Sundarbans

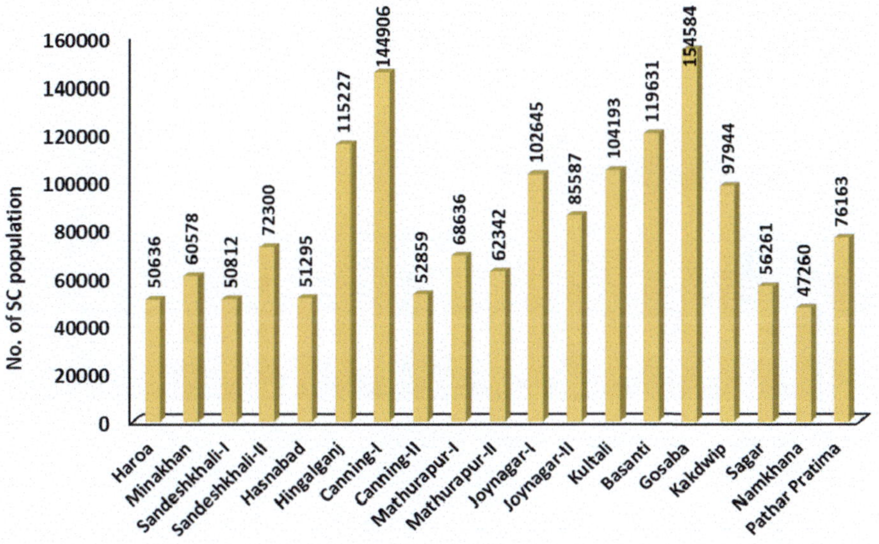

Fig. 1.21 Block-wise Scheduled Caste (SC) population in Indian Sundarbans

magnitude. However, due to the land use changes in this deltaic region for setting up shrimp farm (Fig. 1.23), brick kilns (Fig. 1.24), fish landing stations, tourism units at the cost of mangroves, the near surface atmospheric carbon dioxide level has hiked up to a great extent (Fig. 1.25).

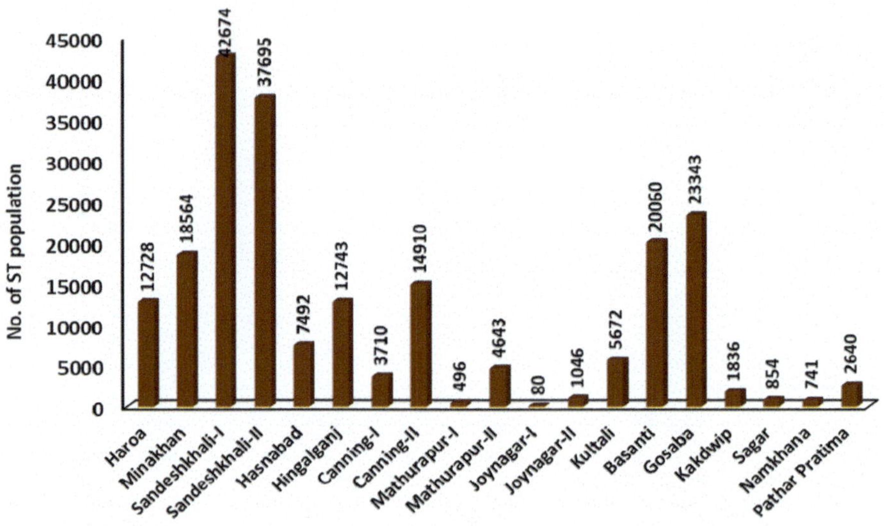

Fig. 1.22 Block-wise Scheduled Tribe (ST) population in Indian Sundarbans

Fig. 1.23 Shrimp farms in Indian Sundarbans at the cost of mangroves

We have used five major indicators to evaluate the footsteps of climate change in the present mangrove dominated delta complex. These are air temperature, near surface atmospheric carbon dioxide, surface water salinity, surface water pH, and Dissolved Oxygen (DO).

Fig. 1.24 Brick Kilns in Indian Sundarbans at the cost of mangroves

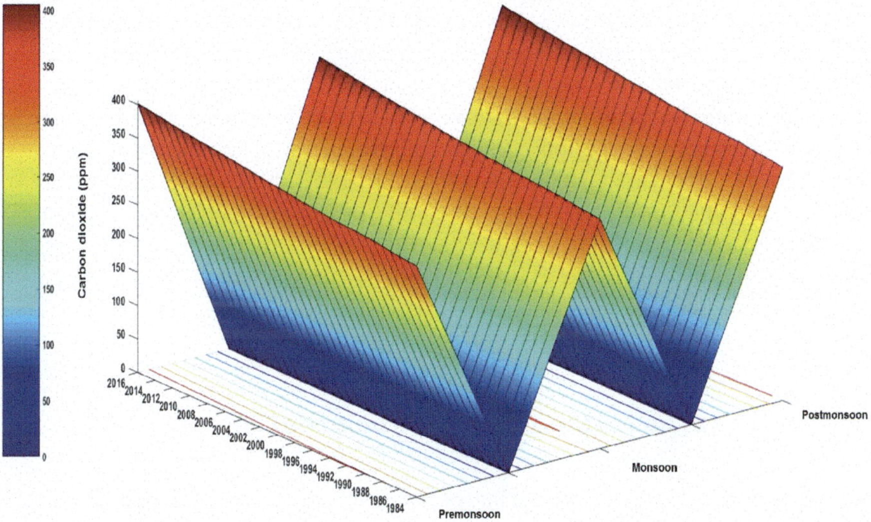

Fig. 1.25 Near surface atmospheric carbon dioxide level at Namkhana in western sector of Indian Sundarbans; *Source* Mitra et al. (2022)

1.3.1 Air Temperature

Carbon dioxide accumulates in the atmosphere mostly due to anthropogenic activities in recent times and causes 'blanket effect' that triggers the rise of air temperature. Earth's temperature has hiked up by an average of 0.08 °C per decade since 1880. The rate of warming since 1981 is more than twice as fast, which is approximately 0.18° C per decade (https://www.climate.gov/news-features/understanding-climate/climate-change-global-temperature). The pulse of rise of air temperature is also felt in the delta region of Indian Sundarbans.

We conducted a study for more than three decades (1984–2022) at Sagar Island (21°38′51.55″N and 88°02′20.97″E) in western Indian Sundarbans, Jharkhali (22°05′52.82″N and 88°41′47.25″E) in central Indian Sundarbans and Jhilla (22°09′51.53″N and 88°57′57.07″E) in eastern Indian Sundarbans during August (monsoon) and observed an increase of 10.24%, 6.73% and 5.83% respectively, which represents a rise of air temperature by 0.087 °C per year, 0.056 °C per year and 0.049 °C per year in these three stations respectively, much more than the above stated global average (Tables 1.12, 1.13 and 1.14 and Fig. 1.26). The carbon foot print seems to be more in this mangrove dominated region possibly due to mass scale destruction of mangroves for promoting shrimp farms that brings short term gain in terms of livelihood of the local population.

Increased air temperature and subsequent effect on water temperature adversely impact the metabolism, growth, and condition index of fish species. Apart from the fishery sector, paddy cultivation is also affected as the optimum temperature for the cultivation ranges between 25 °C to 35 °C and in May 2023, during our field study in Sundarban, we experienced an average temperature around 40 °C (average temperature of 21 days, at 12.00 noon from 3rd May, 2023 to 23rd May, 2023). According to Fahad et al. (2016) the rice yield decreases by 10% for every 1 °C increase in temperature. Thus, it is needless to say that in a scenario of rising air temperature in Sundarbans, the existing livelihoods of Sundarbans people may face serious negative impacts.

1.3.2 Near Surface Atmospheric Carbon Dioxide

The phenomenon of climate change has posed adverse impacts on almost all nations of the World in the domains of agriculture, fishery, biodiversity, human health, economy, and livelihoods. Indian Sundarbans is facing similar situation, which has resulted in the alteration of hydrological parameters of the estuarine water in terms of salinity, pH, dissolved oxygen, nutrient level etc. We measured the carbon dioxide level at 18 m above the ground with a Non-Dispersive Infra-Red (NDIR) gas analyser fitted with a thermometer from 1984–2022 during May (premonsoon season)

Table 1.12 Air temperature (°C) at Sagar Island in western Indian Sundarbans (21°38′51.55″N and 88°02′20.97″E) from 1984–2022 monitored during August (monsoon) in every year

Year	Air temperature (°C)
1984	33.2 ± 0.3
1985	33.6 ± 0.4
1986	33.7 ± 0.3
1987	33.8 ± 0.5
1988	33.9 ± 0.4
1989	34.0 ± 0.5
1990	33.9 ± 0.4
1991	34.2 ± 0.5
1992	34.0 ± 0.4
1993	34.1 ± 0.5
1994	34.3 ± 0.5
1995	34.2 ± 0.5
1996	34.1 ± 0.7
1997	34.2 ± 0.4
1998	34.2 ± 0.6
1999	34.1 ± 0.7
2000	34.6 ± 0.7
2001	34.6 ± 0.8
2002	34.8 ± 0.5
2003	34.8 ± 0.7
2004	34.8 ± 0.8
2005	34.7 ± 0.5
2006	34.9 ± 0.8
2007	34.9 ± 0.7
2008	35.0 ± 0.7
2009	35.1 ± 0.6
2010	34.9 ± 0.7
2011	34.9 ± 0.6
2012	35.1 ± 0.6
2013	35.2 ± 0.6
2014	35.3 ± 0.6
2015	35.7 ± 0.5
2016	35.8 ± 0.7
2017	35.9 ± 0.5
2018	36.2 ± 0.6
2019	36.1 ± 0.7
2020	36.3 ± 0.8
2021	36.4 ± 0.8
2022	36.6 ± 0.8

Table 1.13 Air temperature (°C) at Jharkhali in central Indian Sundarbans (22°05′52.82″N and 88°41′47.25″E) from 1984–2022 monitored during August (monsoon) in every year

Year	Air temperature (°C)
1984	32.7 ± 0.5
1985	33.1 ± 0.5
1986	33.1 ± 0.4
1987	33.2 ± 0.5
1988	33.3 ± 0.4
1989	33.3 ± 0.4
1990	33.5 ± 0.4
1991	33.3 ± 0.4
1992	33.3 ± 0.5
1993	33.6 ± 0.5
1994	33.8 ± 0.7
1995	33.5 ± 0.5
1996	33.3 ± 0.6
1997	33.5 ± 0.6
1998	33.4 ± 0.6
1999	33.4 ± 0.4
2000	33.6 ± 0.4
2001	33.0 ± 0.6
2002	33.6 ± 0.5
2003	33.7 ± 0.7
2004	33.8 ± 0.5
2005	34.0 ± 0.6
2006	34.0 ± 0.7
2007	34.0 ± 0.7
2008	33.9 ± 0.7
2009	34.5 ± 0.4
2010	34.6 ± 0.5
2011	34.9 ± 0.5
2012	34.2 ± 0.6
2013	34.4 ± 0.5
2014	34.9 ± 0.6
2015	35.0 ± 0.6
2016	35.1 ± 0.8
2017	35.0 ± 0.5
2018	34.6 ± 0.8
2019	34.7 ± 0.5
2020	34.2 ± 0.6
2021	34.7 ± 0.6
2022	34.9 ± 0.6

Table 1.14 Air temperature (°C) at Jhilla in eastern Indian Sundarbans (22°09′51.53″N and 88°57′57.07″E) from 1984–2022 monitored during August (monsoon) in every year

Year	Air temperature (°C)
1984	32.6 ± 0.3
1985	32.6 ± 0.3
1986	32.6 ± 0.3
1987	32.9 ± 0.4
1988	33.0 ± 0.3
1989	33.0 ± 0.3
1990	33.0 ± 0.4
1991	33.2 ± 0.4
1992	32.9 ± 0.5
1993	34.3 ± 0.4
1994	33.4 ± 0.3
1995	33.0 ± 0.5
1996	32.9 ± 0.2
1997	33.1 ± 0.3
1998	33.1 ± 0.3
1999	33.2 ± 0.4
2000	33.2 ± 0.3
2001	32.5 ± 0.4
2002	33.4 ± 0.2
2003	33.6 ± 0.1
2004	33.5 ± 0.3
2005	33.5 ± 0.3
2006	33.6 ± 0.4
2007	33.8 ± 0.5
2008	33.5 ± 0.6
2009	34.1 ± 0.5
2010	33.8 ± 0.6
2011	33.7 ± 0.3
2012	33.7 ± 0.4
2013	34.0 ± 0.5
2014	34.1 ± 0.5
2015	34.5 ± 0.5
2016	34.6 ± 0.6
2017	34.7 ± 0.5
2018	34.7 ± 0.4
2019	34.9 ± 0.5
2020	34.8 ± 0.5
2021	34.7 ± 0.6
2022	34.5 ± 0.6

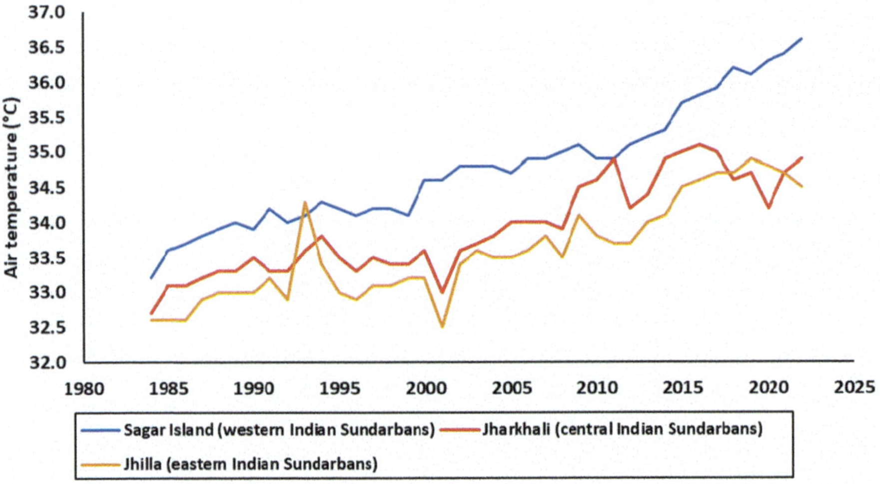

Fig. 1.26 Air temperature (°C) in three selected stations in three different sectors of Indian Sundarbans during monsoon period from 1984 to 2022

in every year at Sagar Island (21°38′51.55″N and 88°02′20.97″E) in the western sector, Thakuran (21°49′43.17″N and 88°33′21.57″E) in the central sector and Jhilla (22°09′51.53″N and 88°57′57.07″E) in the eastern sector of Indian Sundarbans.

We observed increasing trends of this green house gas in all the three sectors except during the COVID-19 lockdown phase in 2020. This may be attributed to complete paralysis of industrial and other anthropogenic activities like fishing, activities of the fish landing station, construction works, brick making in the brick kilns, tourism, religious gatherings etc. that are very relevant in the present study are (Tables 1.15, 1.16 and 1.17).

Based on these ground-zero data, we carried out Non-Linear-Autoregressive model (NAR) to evaluate the situation in the three sectors during 2050 (Figs. 1.27, 1.28 and 1.29) that reveal values of 551 ppm, 504 ppm and 501 ppm in the western, central, and eastern Indian Sundarbans respectively. The significant variations of anthropogenic influences might be the cause behind such spatial variation (Table 1.18).

It has been documented by researchers that under a scenario of elevated carbon dioxide, the concentrations of protein and essential minerals in plants is considerably reduced, which can be a direct adverse impact on human health. Fish diversity, which is the basic foundation of Sundarban fisherman is also negatively affected due to elevated dissolved carbon dioxide in the estuarine water.

Table 1.15 Near Surface Atmospheric Carbon dioxide level (ppm) at Sagar Island in western Indian Sundarbans (21°38′51.55″N and 88°02′20.97″E) from 1984–2022 monitored during May in every year

Year	CO_2 (ppm)
1984	375
1985	372
1986	377
1987	374
1988	378
1989	379
1990	381
1991	380
1992	390
1993	391
1994	396
1995	395
1996	397
1997	401
1998	403
1999	410
2000	404
2001	413
2002	406
2003	398
2004	401
2005	404
2006	407
2007	404
2008	406
2009	401
2010	402
2011	409
2012	407
2013	411
2014	408
2015	412
2016	410
2017	413
2018	414
2019	412
2020	341*

(continued)

Table 1.15 (continued)

Year	CO_2 (ppm)
2021	369
2022	423

* The lowest value of atmospheric carbon dioxide during 2020 is the reflection of COVID-19 lockdown phase (*Vide* Mitra 2023 for further details)

Very recently the term carbon footprint has taken a major share in almost all the sectors of our life and activities. We think it as a proxy that represents the amount of greenhouse gases (GHGs), expressed as CO_2 equivalents, that are emitted directly or indirectly as a result of a specific activity ranging domestic to defense. The activity includes a very wide range like manufacturing, food consumption, life style, transportation, change of land use pattern, livelihoods like agriculture, fishery, aquaculture, animal husbandry, and even war, political, religious and social gatherings etc.

The carbon footprint of an individual is calculated in terms of major categories of consumption: housing, travel, food, products and services. The input per individual includes things like fuel use, calorie consumption, expenditure, distance, and spending behaviour. The appropriate emissions factor takes into account the relevant life cycle as much as possible. On an average, a city-dweller emits four tonnes of carbon dioxide in a year. The realistic figure of carbon footprint for an island dweller of Indian Sundarbans is complicated to calculate as it is a function of his livelihood types, duration of using fossil fuel based transportation, duration of using aerator and diesel operated pump in shrimp farms, engagement in deforestation activity, fishing activity using fuel operated trawlers, running a brick kiln, small scale tourism units etc.

To be honest, the main contributers to carbon footprints in the present region includes mangrove deforestation for setting up shrimp farms, brick kilns, fish landing stations, tourism units. Even the fishing vessels, trawlers and passenger vessels used by the island dwellers to move across the islands or reaching the nearest cities and towns (like Kolkata, Haldia or Howrah) contribute to substantial carbon footprint owing to use of fossil fuels (preferably diesel).

The different activities that contribute to carbon footprint in Indian Sundarbans is not uniform in terms of space and time. There is significant saptial variation *e.g.*, industrial activities are highest in western sector, but in the eastern sector encompassing the Researve Forest (RF), there are no industries. Again, pilgrim related activities are maximum in the western part of Indian Sundarbans due to presence of Kapil Muni temple (22°38′15.35″N and 88°04′30.77″E). This temple is one of the hotspots of Indian pilgrimage, and every year during Makar Sankranti (13–15th January) about 10–15 lakhs people from different corners of the Indian sub-continent gather here to take the Holy bath at the confluence of the Hoogly estuary and Bay of Bengal adjacent to the temple (Fig. 1.30).

Table 1.16 Near Surface Atmospheric Carbon dioxide level (ppm) at Thakuran (21°49′43.17″N and 88°33′21.57″E) in the central sector of Indian Sundarbans from 1984–2022 monitored during May in every year

Year	CO_2 (ppm)
1984	369
1985	371
1986	365
1987	372
1988	370
1989	366
1990	368
1991	371
1992	375
1993	380
1994	383
1995	385
1996	388
1997	390
1998	391
1999	394
2000	396
2001	398
2002	399
2003	401
2004	403
2005	405
2006	407
2007	411
2008	413
2009	418
2010	404
2011	403
2012	409
2013	411
2014	407
2015	412
2016	414
2017	403
2018	409
2019	408
2020	350*

(continued)

Table 1.16 (continued)

Year	CO_2 (ppm)
2021	363
2022	407

* The lowest value of atmospheric carbon dioxide during 2020 is the reflection of COVID -19 lockdown phase (*Vide* Mitra 2023 for further details)

Carbon footprint created by the activities of the island dwellers play a vital role in shrinking the natural resource base of the region and harming the environment. Every human activity which generates carbon foot print poses an adverse impact on the ambient environment. In the present delta complex, activities like mangrove deforestation for setting shrimp farms, brick kilns or tourism units are perhaps the worst as these lead to erosion and eventually increase the degree of vulnerability in terms of lives and livelihoods.

Table 1.18 presents the magtitude of different activities related to carbon footprints in the western, central and eastern sectors of Indian Sundarbans.

1.3.3 Surface Water Salinity

In this section we present evidences that the Indian Sundarbans is experiencing the phenomenon of alteration of salinity as direct impact of climate change at the local scale. Basically, this deltaic complex serves as a natural laboratory to test the impact of climate change owing to the presence of three different sectors having contrasting salinity profiles. Increased melting of the Himalayan glaciers might be the reason of decrease in salinity in the western Indian Sundarbans due to input of fresh water sourced from the Himalayan glaciers through mighty River Ganga that ends up at Bay of Bengal.

In the central part of Indian Sundarbans, a significant increase in aquatic salinity is observed with the passage of time, which might be due to heavy siltation of the Bidyadhari channel since the fifteenth century (Chaudhuri and Choudhury 1994; Mitra et al. 2009; Mitra et al. 2010; Barua et al. 2011; Raha et al. 2012; Bhattacharyya et al. 2013; Mitra and Ghosh 2014; Mitra et al. 2014; Agarwal et al. 2016a; Agarwal et al. 2016b; Chakraborty et al. 2016; Mitra et al. 2016; Banerjee et al. 2017; Ray Chaudhuri et al. 2017; Agarwal and Mitra 2018). This has blocked the fresh water inflow in the Matla estuary in the central Indian Sundarbans due to which salinity is exhibiting an increasing trend since the last few decades.

Table 1.17 Near Surface Atmospheric Carbon dioxide level (ppm) at Jhilla (22°09′51.53″N and 88°57′57.07″E) in the eastern sector of Indian Sundarbans from 1984–2022 monitored during May in every year

Year	CO_2 (ppm)
1984	355
1985	357
1986	355
1987	357
1988	358
1989	356
1990	360
1991	362
1992	365
1993	370
1994	372
1995	375
1996	380
1997	391
1998	394
1999	390
2000	393
2001	394
2002	395
2003	397
2004	401
2005	398
2006	395
2007	402
2008	395
2009	393
2010	403
2011	401
2012	405
2013	403
2014	406
2015	408
2016	410
2017	407
2018	405
2019	407
2020	341*

(continued)

Table 1.17 (continued)

Year	CO_2 (ppm)
2021	352
2022	398

* The lowest value of atmospheric carbon dioxide during 2020 is the reflection of COVID-19 lockdown phase (*Vide* Mitra 2023 for further details)

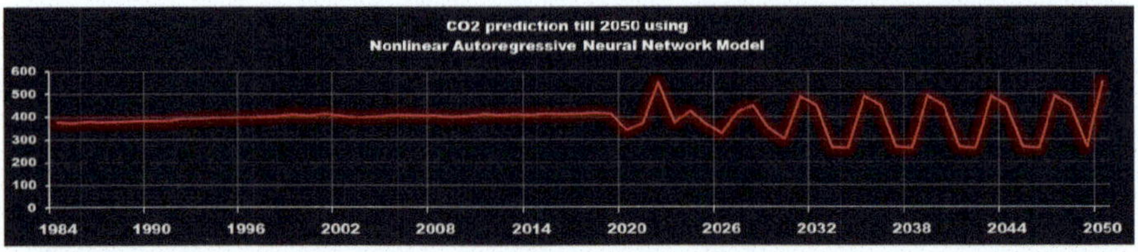

Fig. 1.27 Predicted near surface atmospheric carbon dioxide level for May 2050 in western Indian Sundarbans using nonlinear Autoregressive Neural Network Model; real-time data from 1984–2022 has been used to train the model

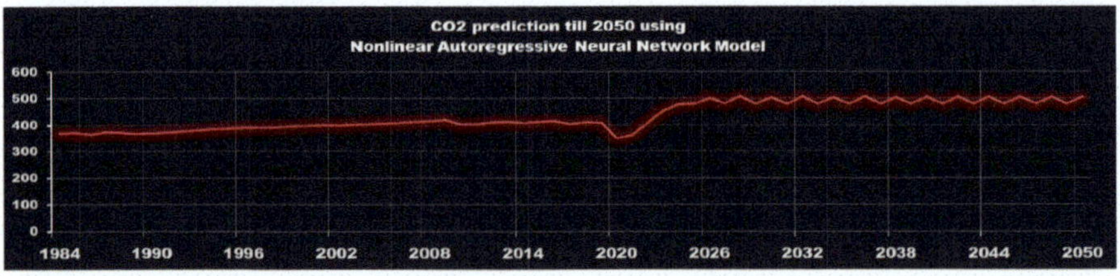

Fig. 1.28 Predicted near surface atmospheric carbon dioxide level for May 2050 in central Indian Sundarbans using nonlinear Autoregressive Neural Network Model; real-time data from 1984–2022 has been used to train the model

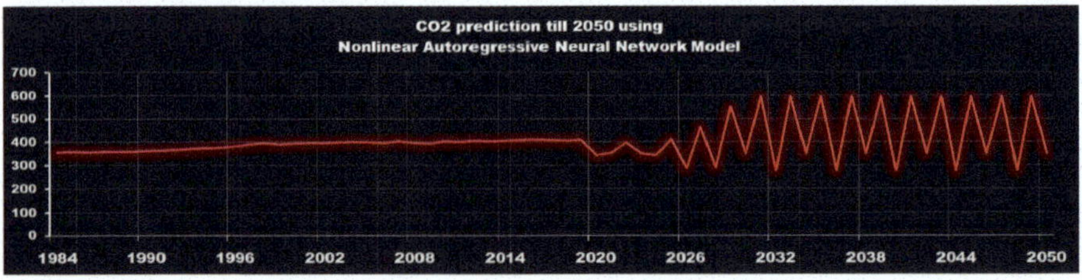

Fig. 1.29 Predicted near surface atmospheric carbon dioxide level for May 2050 in eastern Indian Sundarbans using nonlinear Autoregressive Neural Network Model; real-time data from 1984–2022 has been used to train the model

Table 1.18 Spatial variations of different anthropogenic activities in Indian Sundarbans

Indian Sundarban Sector	Activity									
	Fish landing	Movement of Fishing vessels and trawlers	Passenger vessel related activity	Conditioning of boats, vessels, and trawlers	Shrimp farming	Industrial activity	Brick kilns	Tourism	Pilgrim related activity	Navigation
Western Sector	Maximum	Maximum	Maximum	Maximum	Medium	Maximum	Maximum	Maximum	Maximum	Medium
Central Sector	Less	Less	Maximum	Medium	Maximum	Least	Least	Medium	Less	Least
Eastern Sector	Least	No information	Least	No information	Nil	Nil	Nil	Medium	Nil	Medium

INDEX

- Maximum
- Medium
- Less
- Least
- Nil
- No information

We also carried out the forecasting model using NAR in two stations of the study area namely Kakdwip (21°52′26.50″N and 88°08′04.48″E) and Jharkhali (22°05′52.82″N and 88°41′47.25″E) located in the western and central Indian Sundarbans respectively and the results show salinity levels of 4.31 psu and 30.52 psu in 2050 in the aquatic system of Kakdwip and Jharkhali respectively (Tables 1.19 and 1.20; Figs. 1.31 and 1.32).

Climate change induced alteration of salinity influences the biomass, growth, and carbon storage potential of halophytes. Stunted growth of halophytes is observed in high saline zone due to which the carbon storage efficiency is greatly reduced (Mitra 2013; Mitra 2020; Mitra and Zaman 2021; Mitra et al. 2022; Mitra 2023). The stunted growth of mangroves results in the retardation of ecosystem services like timber, fuel and fodder production, bioremediation, nutrient retention, erosion control etc. The livelihood related income also faces adverse effect due to replacement of commercially important fishes by trash varieties.

Fig. 1.30 Kapil Muni temple at Sagar Island located at the confluence of the Hooghly estuary and Bay of Bengal

The increasing trend of salinity in Sundarban estuaries (preferably in the central sector of the delta) not only increase the salt content of the soil and affect the growth of food crops and vegetables, but also invite trash variety of fish species in place of commercially important fish species like *Liza parsia*, *Liza tade* etc. (Mitra 2013). The Indian shad, *Tenualosa ilisha* breed only in freshwater and hence their migratory route gets altered with the intrusion of saline water in the estuaries of Sundarbans. Such intrusion is natural due to sea level rise and cyclones, when the saline water from the adjacent Bay of Bengal finds the way in the inshore regions of the Sundarban delta complex.

Table 1.19 Surface Water Salinity (psu) at Kakdwip in western Indian Sundarbans (21°52′26.50″N and 88°08′04.48″E) from 1984–2022 monitored during May in every year; real-time data from 1984–2022 has been used to forecast salinity till 2050

Year	Salinity (psu)
1984	11.68
1985	11.04
1986	10.87
1987	10.99
1988	11.55
1989	11.02
1990	10.85
1991	10.32
1992	10.07
1993	9.97
1994	9.45
1995	9.63
1996	8.86
1997	8.31
1998	7.7
1999	7.95
2000	7.04
2001	7.43
2002	6.51
2003	6.25
2004	6.2
2005	6.39
2006	5.77
2007	5.17
2008	5.07
2009	11.04*
2010	5.00
2011	4.66
2012	5.01
2013	4.93
2014	4.26
2015	5.07
2016	4.09
2017	4.11
2018	4.17
2019	3.99
2020	4.19
2021	4.19

(continued)

Table 1.19 (continued)

Year	Salinity (psu)
2022	4.25
2023	4.26
2024	4.28
2025	4.29
2026	4.30
2027	4.30
2028	4.31
2029	4.31
2030	4.31
2031	4.31
2032	4.31
2033	4.31
2034	4.31
2035	4.31
2036	4.31
2037	4.31
2038	4.31
2039	4.31
2040	4.31
2041	4.31
2042	4.31
2043	4.31
2044	4.31
2045	4.31
2046	4.31
2047	4.31
2048	4.31
2049	4.31
2050	4.31

* The sudden hike of surface water salinity during 2009 is the impact of Aila (a super cyclone), which resulted in the intrusion of sea water from Bay of Bengal in the estuarine system

1.3.4 Surface Water pH

Increased carbon dioxide level in the atmosphere is the basic root of acidification of marine and estuarine waters. Very few studies have been carried out on this domain in the present geographical locale (Agarwal et al. 2019; Mitra and Zaman 2021), but

Table 1.20 Surface Water Salinity (psu) at Jharkhali in central Indian Sundarbans (22°05′52.82″N and 88°41′47.25″E) from 1984–2022 monitored during May in every year; real-time data from 1984–2022 has been used to forecast salinity till 2050

Year	Salinity (psu)
1984	22.07
1985	23.59
1986	24.34
1987	24.22
1988	25.04
1989	25.69
1990	25.84
1991	25.01
1992	24.72
1993	25.60
1994	26.00
1995	26.04
1996	26.08
1997	26.83
1998	26.81
1999	26.75
2000	26.98
2001	27.04
2002	27.13
2003	27.45
2004	27.2
2005	27.18
2006	27.86
2007	27.69
2008	28.34
2009	31.68*
2010	29.87
2011	30.13
2012	30.08
2013	30.46
2014	30.73
2015	30.81
2016	31.09
2017	30.8
2018	30.96
2019	31.33
2020	30.59
2021	31.23

(continued)

Table 1.20 (continued)

Year	Salinity (psu)
2022	30.36
2023	30.97
2024	30.61
2025	31.15
2026	30.57
2027	31.17
2028	30.48
2029	31.09
2030	30.53
2031	31.12
2032	30.53
2033	31.13
2034	30.52
2035	31.12
2036	30.52
2037	31.12
2038	30.52
2039	31.12
2040	30.52
2041	31.12
2042	30.52
2043	31.12
2044	30.52
2045	31.12
2046	30.52
2047	31.12
2048	30.52
2049	31.12
2050	30.52

* The highest value of surface water salinity during 2009 is the impact of Aila (a super cyclone), which resulted in the intrusion of sea water from Bay of Bengal in the estuarine system

the issue is extremely important as lowering of pH may disrupt the estuarine biodiversity preferably the molluscan community. We conducted a study for more than three decades (1984–2022) at Bali Island (22°04′35.17″N and 88°44′55.70″E) in central Indian Sundarbans during August (monsoon) and December (postmonsoon) and observed the decrease of 1.32% and 1.08% in monsoon and postmonsoon respectively, which represents a fall of pH by 0.0028 unit per year and 0.0023 unit per year

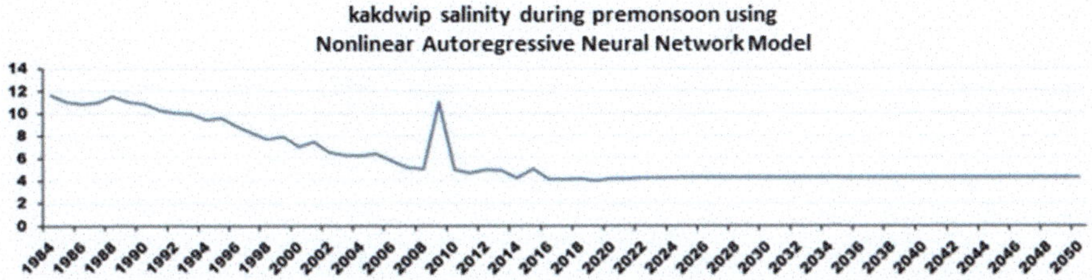

Fig. 1.31 Predicted surface water salinity level (ppm) for May 2050 at Kakdwip (in western Indian Sundarbans) using Nonlinear Autoregressive Neural Network Model; real-time data from 1984–2022 has been used to train the model

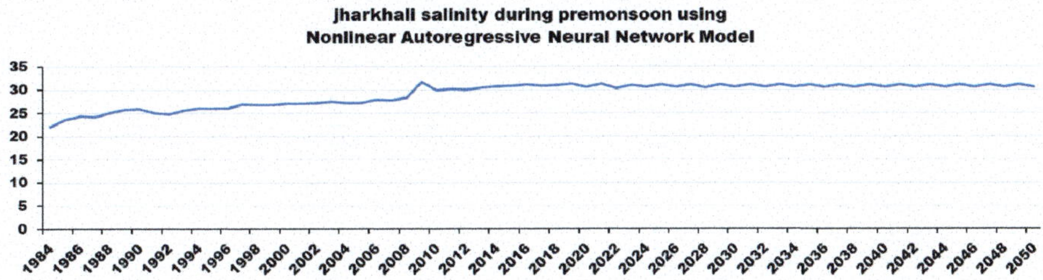

Fig. 1.32 Predicted surface water salinity level (ppm) for May 2050 in central Indian Sundarbans using nonlinear Autoregressive Neural Network Model; real-time data from 1984–2022 has been used to train the model

in these two seasons (Tables 1.21 and 1.22), compared to the global data. According to the estimated global mean surface sea water pH, there has been a decrease in pH since 1985 of 0.0016 pH units per year, with an error on each yearly value of 0.0006 (http://marine.copernicus.eu/).

Table 1.21 Surface Water pH at Bali Island in central Indian Sundarbans (22°04′35.17″N and 88°44′55.70″E) from 1984–2022 monitored during August (monsoon) in every year

Year	pH
1984	8.31
1985	8.32
1986	8.32
1987	8.31
1988	8.31
1989	8.33
1990	8.3
1991	8.32
1992	8.31
1993	8.32
1994	8.32
1995	8.3
1996	8.33
1997	8.32
1998	8.32
1999	8.31
2000	8.3
2001	8.31
2002	8.31
2003	8.31
2004	8.31
2005	8.31
2006	8.3
2007	8.3
2008	8.29
2009	8.3
2010	8.29
2011	8.29
2012	8.28
2013	8.27
2014	8.26
2015	8.24
2016	8.23
2017	8.22
2018	8.21
2019	8.2
2020	8.2
2021	8.19
2022	8.20

Table 1.22 Surface Water pH at Bali Island in central Indian Sundarbans (22°04′35.17″N and 88°44′55.70″E) from 1984–2022 monitored during December (postmonsoon) in every year

Year	pH
1984	8.33
1985	8.34
1986	8.33
1987	8.33
1988	8.34
1989	8.33
1990	8.33
1991	8.33
1992	8.32
1993	8.34
1994	8.33
1995	8.33
1996	8.32
1997	8.33
1998	8.33
1999	8.29
2000	8.32
2001	8.32
2002	8.33
2003	8.31
2004	8.31
2005	8.33
2006	8.32
2007	8.31
2008	8.31
2009	8.29
2010	8.31
2011	8.31
2012	8.29
2013	8.28
2014	8.28
2015	8.26
2016	8.24
2017	8.24
2018	8.23
2019	8.23
2020	8.22
2021	8.23
2022	8.24

Based on these ground-zero data, application of Non-Linear-Autoregressive model (NAR) to evaluate the situation in these two seasons exhibit a further drop in pH values (Tables 1.23 and 1.24) that may worsen the situation on molluscan species. The bivalves and gastropods are not only the important components of the deltaic food chain, but majority of them are the sources of lime that serve as the base of livelihoods for many of the island dwellers.

Table 1.23 Outputs of the Non-Linear-Autoregressive model (NAR) based on the ground zero data on Surface Water pH at Bali Island in central Indian Sundarbans (22°04′35.17″N and 88°44′55.70″E) from 1984–2022 monitored during August (monsoon) in every year

Year	pH
2023	8.19
2024	8.20
2025	8.19
2026	8.20
2027	8.19
2028	8.20
2029	8.19
2030	8.20
2031	8.19
2032	8.20
2033	8.19
2034	8.20
2035	8.19
2036	8.20
2037	8.19
2038	8.20
2039	8.19
2040	8.20
2041	8.19
2042	8.20
2043	8.19
2044	8.20
2045	8.19
2046	8.20
2047	8.19
2048	8.20
2049	8.19
2050	8.18

Table 1.24 Outputs of the Non-Linear-Autoregressive model (NAR) based on the ground zero data on Surface Water pH at Bali Island in central Indian Sundarbans (22°04′35.17″N and 88°44′55.70″E) from 1984–2022 monitored during December (postmonsoon) in every year

Year	pH
2023	8.24
2024	8.23
2025	8.23
2026	8.23
2027	8.23
2028	8.23
2029	8.23
2030	8.23
2031	8.23
2032	8.23
2033	8.23
2034	8.23
2035	8.23
2036	8.23
2037	8.23
2038	8.23
2039	8.23
2040	8.23
2041	8.23
2042	8.23
2043	8.23
2044	8.23
2045	8.23
2046	8.23
2047	8.23
2048	8.23
2049	8.23
2050	8.23

1.3.5 Dissolved Oxygen (DO)

Climate change is known to cause deoxygenation in the open ocean, but its effects on eutrophic and seasonally hypoxic estuaries and coastal oceans are less clear (Ni et al. 2019). However, in case of estuaries, an inverse relationship is often witnessed between dissolved oxygen and water temperature. As the temperature of the water increases, dissolved oxygen levels decrease, although other factors like nutrient levels, pathogens, effluents, and industrial wastes play significant role in the dynamics of dissolved oxygen in the estuarine water (Amin et al. 2015).

We conducted a study for more than three decades (1984–2022) at Sagar Island (21°38′51.55″N and 88°02′20.97″E) in western Indian Sundarbans, Thakuran (21°49′43.17″N and 88°33′21.57″E) in central Indian Sundarbans and Jhilla (22°09′51.53″N and 88°57′57.07″E) in eastern Indian Sundarbans during August (monsoon) and observe a fall by 3.03% and 24.47% in Sagar Island and Thakuran respectively, which represents a reduction of the DO level by 0.0054 ppm per year and 0.0418 ppm per year in these two stations respectively. In Jhilla, DO increases 2.65% representing a rise of 0.0046 ppm per year (Table 1.25 and Fig. 1.33).

DO is the most important indicator of aquatic health and acts as the life line of survival of aquatic organisms. It is assessed by the amount of oxygen that is absorbed in given mass of water body. Oxygen has a limited solubility range (6–10 mg/L) in tropical brackish water. However, the solubility is a function of salinity and decreases considerably with the increase of salinity value. Hence, the trend of increasing salinity in the estuarine water of Sundarbans will have drastic impact on the fish communities in terms of their survival, growth, and composition.

1.4 Take Home Messages

(A) Sundarban in the Indian part sustains a wide spectrum of flora encompassing 37 genera of bacteria, 5 genera of fungi, 1 genus of virus, 270 species of algae, 71 species of diatoms, 167 species of lichens and 178 species of mangrove and mangrove associate species. The faunal community consists of zooplankton dominated by 52 species of copepods in the estuarine water. Apart from zooplankton, 67 species of protozoa, 177 species of mollusca, 57 species of polychaetes, 2 species of xiphosurans, 329 species of crustaceans, 114 species of spiders, 121 species of mites, 497 species of insects, 285 species of fishes, 11 species of amphibia, 71 species of reptiles, 250 species of aves, and 47 species of mammals have been recorded from the region.

(B) The Indian Sundarban occupies 38% share of the total Sundarbans and the remaining 62% lies within the Bangladesh territory that lies to the east of the Indian part of Sundarbans. The region is noted for the presence of seven major estuaries, but more important is the significant variation of water quality during two tidal phases that governs the biotic community in the delta.

(C) The term carbon footprint has become much common in all the sectors of our life and activities. The realistic figure of carbon footprint for an island dweller of Indian Sundarbans is complicated to calculate as it is a function of his livelihood types, duration of using fossil fuel based transportation, duration of using aerator and diesel operated pump in shrimp farms, engagement in deforestation activity, fishing activity using fuel operated trawlers, running a brick kiln, small scale tourism units etc. To be honest, the main contributers to carbon footprints in the present region includes mangrove deforestation for setting up shrimp farms, brick kilns, fish landing stations, tourism units. Even the fishing vessels, trawlers and passenger vessels used by the island dwellers to

Table 1.25 Yearly variation of dissolved oxygen (ppm) in three stations of Indian Sundarbans from 1984–2022 monitored during August (monsoon) in every year

Year	Sagar Island (21°38′51.55″N and 88°02′20.97″E)	Thakuran (21°49′43.17″N and 88°33′21.57″E)	Jhilla (22°09′51.53″N and 88°57′57.07″E)
1984	6.94	6.66	6.79
1985	6.91	6.24	6.72
1986	6.45	6.21	6.32
1987	6.24	5.89	6.01
1988	6.13	5.84	6.03
1989	6.11	5.78	6.01
1990	6.09	5.69	5.99
1991	6.11	5.67	5.88
1992	6.65	5.74	6.12
1993	6.13	5.81	6.02
1994	6.23	5.8	6.01
1995	6.11	5.76	5.97
1996	6.12	5.73	5.87
1997	6.34	5.69	6.03
1998	6.79	5.56	6.31
1999	6.33	5.43	5.99
2000	6.27	5.65	5.78
2001	6.3	5.34	5.96
2002	6.29	5.13	6.08
2003	6.12	5.11	5.99
2004	6.19	5.89	6.03
2005	6.38	5.78	6.09
2006	6.49	5.89	6.23
2007	6.73	5.76	6.34
2008	6.23	5.56	5.99
2009	6.9	7.36	6.17
2010	6.69	6.66	6.91
2011	6.19	5.57	5.87
2012	6.85	5.25	6.56
2013	6.15	5.67	5.89
2014	6.89	5.14	6.34
2015	6.65	5.33	6.02
2016	6.33	5.65	6.00
2017	6.49	5.22	6.18
2018	6.56	5.17	6.70

(continued)

Table 1.25 (continued)

Year	Sagar Island (21°38′51.55″N and 88°02′20.97″E)	Thakuran (21°49′43.17″N and 88°33′21.57″E)	Jhilla (22°09′51.53″N and 88°57′57.07″E)
2019	6.38	5.13	6.83
2020	6.71	5.10	6.19
2021	6.55	5.09	6.94
2022	6.73	5.03	6.97

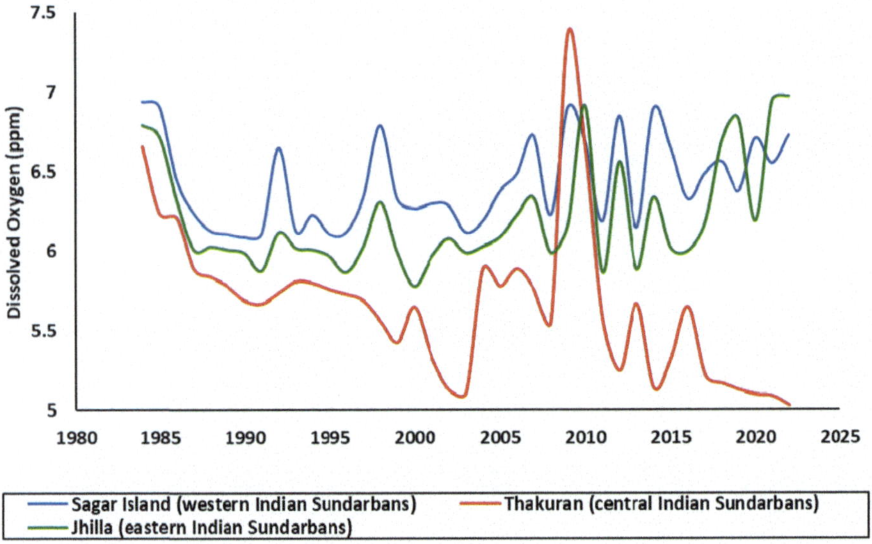

Fig. 1.33 Dissolved Oxygen (ppm) in three selected stations in three different sectors of Indian Sundarbans during monsoon period from 1984 to 2022

move across the islands or reaching the nearest cities and towns (like Kolkata, Haldia or Howrah) contribute to substantial carbon footprint owing to use of fossil fuels (preferably diesel). The carbon footprints generated by the island dwellers leave prominent signature in the climatic profile at local scale.

(D) The different activities that contribute to carbon foot print in Indian Sundarbans is not uniform in terms of space and time. There is significant saptial variation e.g., industrial activities are highest in western sector, but in the eastern sector encompassing the Researve Forest (RF), there are no industries. Again, pilgrim related activities are maximum in the western part of Indian Sundarbans due to presence of Kapil Muni temple (22°38′15.35″N and 88°04′30.77″E). This temple is one of the hotspots of Indian pilgrimage, and every year during Makar Sankranti (13–15th January) about 10–15 lakhs people from different corners of the Indian sub-continent gather here to take the Holy bath at the confluence of the Hooghly estuary and Bay of Bengal adjacent to the temple.

References

Agarwal SK, Mitra A (2018) Salinity: a primary growth driver of mangrove flora. Curr Trend for Res 2:1–9

Agarwal S, Pramanick P, Zaman S, Roy A, Mitra A (2016a) Status of stored carbon in mangroves of lower Gangetic delta. Glob J Eng Res 7–16

Agarwal S, Zaman S, Biswas S, Pal N, Pramanick P, Mitra A (2016b) Spatial variation of mangrove seedling carbon with respect to salinity: a case study with *Bruguiera gymnorrhiza* seedling. Int J Adv Res Biol Sci 3(8):7–12

Agarwal S, Fazli P, Zaman S, Pramanick P, Mitra A (2019) Seasonal variability of acidification in major estuaries of Indian Sundarbans. Glob J Eng Res 6(4):493–498

Amin G, Biswas S, Zaman S, Pramanick P, Trivedi S, Mitra A (2015) Prediction of dissolved oxygen in Indian Sundarbans; vision 2050. Int Adv Res J Sci Eng Technol 2(12):31–33

Banerjee K, Gatti RC, Mitra A (2017) Climate change-induced salinity variation impacts on a stenoecious mangrove species in the Indian Sundarbans. Ambio (Springer) 46:492–499

Banerjee K, Vyas P, Chowdhury R, Mallik A. Mitra A (2010) The affects of salinity on the mangrove growth in the lower Gangetic delta. J Indian Ocean Stud 18(3):389–397

Barua P, Mitra A, Banerjee K, Chowdhury MSN (2011) Seasonal variation of heavy metals accumulation in water and oyster (*Saccostrea cucullata*) inhabiting central and western sector of Indian Sundarbans. Environ Res J 5:121–130. https://doi.org/10.3923/erj.2011.121.130

Bhattacharyya SB, Roychowdhury G, Zaman S, Raha AK, Chakraborty S, Bhattacharjee AK, Mitra A (2013) Bioaccumulation of heavy metals in Indian white shrimp (*Fenneropenaeus indicus*: a time series analysis). Int J Life Sci Biotechnol Pharma Res 2(2):97–113

Chakraborty S, Rudra T, Guha A, Ray A, Pal N, Mitra A (2016) Spatial variation of heavy metals in *Tenualosa ilisha* muscle: a case study from the lower Gangetic delta and coastal West Bengal. Int J Innov Sci Eng Technol 3(4):1–14

Chaudhuri AB, Choudhury A (1994) Mangroves of the Sundarbans. Volume I: India. IUCN—The World Conservation Union, p 165

Danda AA, Joshi AK, Ghosh A, Saha R (eds) (2017) State of art report on biodiversity in Indian Sundarbans. World Wide Fund for Nature-India, New Delhi

Dhar I, Sengupta G, Biswas S, Sinha M, Mitra A (2021) Salinity: a major environmental factor in sustainability of the blue carbon. J Mech Continua Math Sci 16(11):34–42. https://doi.org/10.26782/jmcms.2021.11.00004

Fahad S, Hussain S, Saud S, Hassan S, Ihsan Z, Shah AN, Wu C, Yousaf M, Nasim W, Alharby H, Alghabari F, Huang J (2016) Exogenously applied plant growth regulators enhance the morpho-physiological growth and yield of rice under high temperature. Front Plant Sci 7:1250

Guha T, Mitra A (2020) Salinity—a crucial factor in ecological sustainability for Sundarbans Mangrove ecosystem. In: Mitra A, Calma MM, Chakrabarty SP (eds) Proceedings: natural resources and their ecosystem services. HSRA Publication, pp 27–35. ISBN 978-81-947216-7-3

http://marine.copernicus.eu/

https://www.climate.gov/news-features/understanding-climate/climate-change-global-temperature

https://www.sundarbanaffairswb.in/home/block

Jana HK, Mitra A, Zaman S, Bose R, Raha AK (2014) Will *Avicennia alba* thrive in climate change induced salinity rise? Int J Sci Res 3(2):459–461

Mitra A (2013) Sensitivity of mangrove ecosystem to changing climate, vol XIX. Springer, New Delhi, Heidelberg, New York, Dordrecht, London, p 323. ISBN-10: 8132215087; ISBN-13: 978-8132215080. ISBN 978-81-322-1509-7. https://doi.org/10.1007/978-81-322-1509-7

Mitra A (2018) Can species serve as proxy to climate change induced salinity alteration? J Mar Bio Aqua 1–3

Mitra A (2020) Mangrove forests in India: exploring ecosystem services, vol XV. Springer, p 361. ISBN 978-3-030-20595-9. https://doi.org/10.1007/978-3-030-20595-9

Mitra A (2023) Impact of COVID-19 lockdown on environmental health: exploring the situation of the lower Gangetic Delta, vol XIII. Springer, p 310. ISBN 978-3-031-27242-4. https://doi.org/10.1007/978-3-031-27242-4

Mitra A, Ghosh R (2014) Bioaccumulation pattern of heavy metals in commercially important fishes on and around Indian Sundarbans. Glob J Anim Sci Res 2(1):33–45

Mitra A, Zaman S (2021) Estuarine acidification: exploring the situation of mangrove dominated Indian Sundarban estuaries, vol XII. Springer, p 402. ISBN 978-3-030-84792-0. https://doi.org/10.1007/978-3-030-84792-0

Mitra A, Banerjee K, Sinha S (2011) Shrimp tissue quality in the lower Gangetic delta at the apex of Bay of Bengal. Toxicol Environ Chem 93(3):565–574

Mitra A, Chowdhury R, Sengupta K, Banerjee K (2010) Impact of salinity on mangroves of Indian Sundarbans. J Coast Conserv 1(1):71–82

Mitra A, Gangopadhyay A, Dube A, Andre CKS, Banerjee K (2009) Observed changes in water mass properties in the Indian Sundarbans (Northwestern Bay of Bengal) during 1980–2007. Curr Sci 97(10):1445–1452

Mitra A, Saha A, Pal N, Chakraborty A, Mitra A, Trivedi S, Zaman S (2016) Estimation of stored carbon in *Sonneratia apetala* seedlings collected from Indian Sundarbans. Indian J Mar Sci 45(11):1598–1602

Mitra A, Zaman S, Bhattacharyya SB (2014) Heavy metal pollution in the lower gangetic mangrove ecosystem. In: Water insecurity: a social dilemma (community, environment and disaster risk management), vol 13. Emerald Group Publishing Limited, Bingley, pp 97–113. https://doi.org/10.1108/S2040-7262(2013)0000013011

Mitra A, Zaman S, Pramanick P (2022) Blue economy in Indian Sundarbans: exploring livelihood opportunities, vol XIV. Springer, p 403. ISBN 978-3-031-07908-5. https://doi.org/10.1007/978-3-031-07908-5

Ni W, Li M, Ross AC, Najjar RG (2019) Large projected decline in dissolved oxygen in a eutrophic estuary due to climate change. J Geophys Res Oceans 124:8271–8289. https://doi.org/10.1029/2019JC015274

Raha AK, Das S, Banerjee K, Mitra A (2012) Climate change impacts on Indian Sunderbans: a time series analysis (1924–2008). Biodivers Conserv (SPRINGER). https://doi.org/10.1007/s10531-012-0260-z)

Ray Chaudhuri T, Dutta P, Zaman S, Mitra A (2017) Status of edible fishes of lower Gangetic Delta in terms of heavy metals. Int J Environ Sci Nat Resour 5(3):555661. https://doi.org/10.19080/IJESNR.2017.05.555661

Sengupta T, Mitra A (2020) Spatial variation of the microbial diversity in the mangrove dominated Sundarban Forest of India. Chapter 14, In: De Mandal S, Bhatt P (eds) Recent advancements in microbial diversity. Academic Press, ELSEVIER, pp 333–350. ISBN 978-0-12-821265-3

Trivedi S, Zaman S, Ray Chaudhuri T, Pramanick P, Fazli P, Amin G, Mitra A (2016) Inter-annual variation of salinity in Indian Sundarbans. Indian J Geo-Mar Sci 45(3):410–415

Chapter 2
Traditional Livelihoods in Sundarban Delta

Contents

Agriculture and fishing related activities are the two major pillars of livelihood in Indian Sundarbans although animal husbandry occupies a considerable share in this domain. These pillars are primarily regulated by climatic conditions at local scale. Due to cyclonic depressions coupled with sea level rise, intrusion of saline water frequently occurs in the coastal villages and islands of Sundarbans that are surrounded by saline and brackish water (Mitra 2013). Even the embankments cannot prevent the entry of saline water from the off-shore region into the brackish and freshwater ecosystem of the islands. In most of the cases, the embankment gets damaged and eroded (Fig. 2.1) due to which the intrusion of saline water cannot be prevented (Mitra et al. 2022).

Major crops like paddy and several vegetables face considerable damage as they cannot withstand saline water. Crop growing is a common practice in the freshwater zone of the island, which also get damaged due to osmotic disbalance that occurs through the mixing of freshwater with the saline water from the adjacent estuaries and coastal regions. We have carried out a detailed study during three major cyclonic depressions that hit Indian Sundarbans during 2009, 2020 and 2021. The study was conducted in 12 stations in and around Indian Sundarbans (Fig. 2.2).

The significant changes in the water mass properties during the super cyclone Aila, Amphan and Yash that devastated the deltaic region of Indian Sundarbans during 2009, 2020 and 2021 respectively are highlighted in Tables 2.1, 2.2 and 2.3.

A. Mitra et al., *Climate Resilient Innovative Livelihoods in Indian Sundarban Delta*,
https://doi.org/10.1007/978-3-031-42633-9_2

Fig. 2.1 Damaged embankment due to wave action is a common feature in Indian Sundarbans

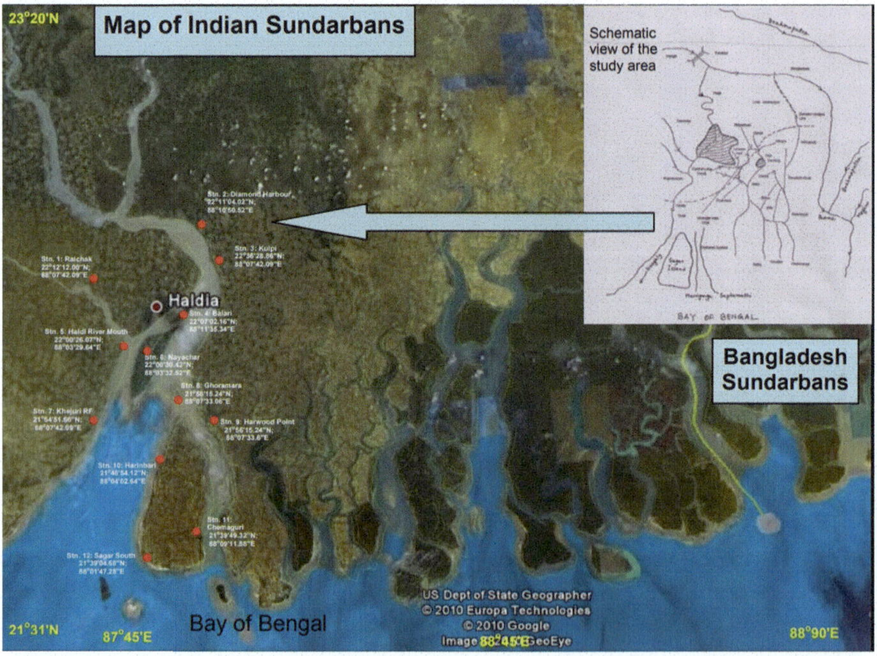

Fig. 2.2 Map showing the location of 12 stations in and around Indian Sundarbans where the effects of three super cyclones were conducted

Table 2.1 Variations of hydrological parameters during different phases of Aila

Stations	Surface water salinity (Phase A)	Surface water salinity (Phase B)	Surface water salinity (Phase C)	D.O (Phase A)	D.O (Phase B)	D.O (Phase C)
Stn. 1	3.39 ± 0.01	3.98 ± 0.01	3.42 ± 0.01	5.69 ± 0.01	4.96 ± 0.02	5.21 ± 0.01
Stn. 2	4.86 ± 0.02	5.91 ± 0.02	4.94 ± 0.01	5.40 ± 0.01	4.98 ± 0.01	5.19 ± 0.02
Stn. 3	6.08 ± 0.02	7.51 ± 0.02	6.16 ± 0.02	6.61 ± 0.02	6.00 ± 0.01	6.49 ± 0.01
Stn. 4	13.14 ± 0.01	16.30 ± 0.01	13.96 ± 0.02	6.53 ± 0.02	5.87 ± 0.02	6.08 ± 0.01
Stn. 5	12.03 ± 0.02	14.94 ± 0.01	12.75 ± 0.03	4.77 ± 0.02	4.72 ± 0.02	5.29 ± 0.02
Stn. 6	13.82 ± 0.02	17.20 ± 0.01	13.99 ± 0.01	4.88 ± 0.03	4.65 ± 0.02	5.09 ± 0.03
Stn. 7	15.52 ± 0.03	19.38 ± 0.01	15.97 ± 0.02	4.80 ± 0.01	4.62 ± 0.01	5.30 ± 0.02
Stn. 8	14.41 ± 0.01	18.02 ± 0.01	14.85 ± 0.01	4.73 ± 0.01	4.40 ± 0.02	5.00 ± 0.02
Stn. 9	17.20 ± 0.02	21.56 ± 0.02	17.93 ± 0.01	4.57 ± 0.01	4.51 ± 0.01	5.97 ± 0.01
Stn. 10	19.31 ± 0.02	24.28 ± 0.02	20.03 ± 0.01	4.65 ± 0.01	4.57 ± 0.02	5.03 ± 0.01
Stn. 11	20.56 ± 0.01	25.90 ± 0.02	20.97 ± 0.02	5.02 ± 0.02	4.97 ± 0.01	5.17 ± 0.02
Stn. 12	23.05 ± 0.02	29.75 ± 0.02	24.64 ± 0.01	5.00 ± 0.02	4.89 ± 0.03	5.25 ± 0.03

Phase A = pre-Aila period (20.05.2009), Phase B = Aila phase (29.05.2009), Phase C = post-Aila phase (06.06.2009); Units of surface water salinity and DO are ppt and ppm respectively

Table 2.2 Variations of hydrological parameters during different phases of Amphan

Stations	Surface water salinity (Phase A)	Surface water salinity (Phase B)	Surface water salinity (Phase C)	D.O (Phase A)	D.O (Phase B)	D.O (Phase C)
Stn. 1	3.37 ± 0.01	5.00 ± 0.01	3.40 ± 0.01	5.72 ± 0.01	4.93 ± 0.02	5.23 ± 0.01
Stn. 2	4.84 ± 0.01	6.92 ± 0.01	4.90 ± 0.01	5.44 ± 0.01	4.94 ± 0.01	5.22 ± 0.02
Stn. 3	6.06 ± 0.02	8.54 ± 0.02	6.13 ± 0.01	6.66 ± 0.02	5.91 ± 0.01	6.53 ± 0.01
Stn. 4	13.12 ± 0.01	17.28 ± 0.01	13.91 ± 0.02	6.55 ± 0.02	5.80 ± 0.02	6.10 ± 0.01
Stn. 5	12.01 ± 0.01	15.95 ± 0.01	12.71 ± 0.02	4.79 ± 0.02	4.68 ± 0.02	5.31 ± 0.02
Stn. 6	13.80 ± 0.02	18.21 ± 0.02	13.97 ± 0.02	4.92 ± 0.03	4.61 ± 0.02	5.13 ± 0.03
Stn. 7	15.50 ± 0.02	20.42 ± 0.01	15.95 ± 0.01	4.86 ± 0.01	4.58 ± 0.01	5.33 ± 0.02
Stn. 8	14.39 ± 0.01	19.02 ± 0.01	14.84 ± 0.01	4.77 ± 0.01	4.35 ± 0.02	5.05 ± 0.02
Stn. 9	17.18 ± 0.01	22.55 ± 0.03	17.89 ± 0.02	4.63 ± 0.01	4.47 ± 0.01	6.01 ± 0.01
Stn. 10	19.29 ± 0.02	25.27 ± 0.02	19.99 ± 0.01	4.69 ± 0.01	4.52 ± 0.02	5.05 ± 0.01
Stn. 11	20.54 ± 0.01	26.92 ± 0.02	20.93 ± 0.02	5.09 ± 0.02	4.94 ± 0.01	5.19 ± 0.02
Stn. 12	23.03 ± 0.02	30.74 ± 0.02	24.59 ± 0.01	5.06 ± 0.02	4.81 ± 0.03	5.28 ± 0.03

Phase A = pre-Amphan period (13.05.2020), Phase B = Amphan phase (20.05.2020), Phase C = post-Amphan phase (04.06.2020); Units of surface water salinity and DO are ppt and ppm respectively

Table 2.3 Variations of hydrological parameters during different phases of Yash

Stations	Surface water salinity (Phase A)	Surface water salinity (Phase B)	Surface water salinity (Phase C)	D.O (Phase A)	D.O (Phase B)	D.O (Phase C)
Stn. 1	3.35 ± 0.01	4.01 ± 0.01	3.38 ± 0.01	5.77 ± 0.01	4.90 ± 0.02	5.26 ± 0.01
Stn. 2	4.80 ± 0.01	5.94 ± 0.02	4.87 ± 0.01	5.48 ± 0.01	4.91 ± 0.01	5.24 ± 0.02
Stn. 3	6.00 ± 0.02	7.53 ± 0.02	6.10 ± 0.01	6.69 ± 0.02	5.88 ± 0.01	6.55 ± 0.01
Stn. 4	13.10 ± 0.01	16.33 ± 0.01	13.88 ± 0.02	6.57 ± 0.02	5.74 ± 0.02	6.13 ± 0.01
Stn. 5	11.98 ± 0.01	14.97 ± 0.01	12.68 ± 0.02	4.83 ± 0.02	4.61 ± 0.02	5.35 ± 0.02
Stn. 6	13.76 ± 0.02	17.23 ± 0.01	13.94 ± 0.02	4.97 ± 0.03	4.58 ± 0.02	5.16 ± 0.03
Stn. 7	15.47 ± 0.02	19.41 ± 0.01	15.90 ± 0.01	4.92 ± 0.01	4.52 ± 0.01	5.37 ± 0.02
Stn. 8	14.33 ± 0.01	18.05 ± 0.01	14.81 ± 0.01	4.83 ± 0.01	4.31 ± 0.02	5.09 ± 0.02
Stn. 9	17.15 ± 0.01	21.58 ± 0.02	17.86 ± 0.02	4.68 ± 0.01	4.43 ± 0.01	6.04 ± 0.01
Stn. 10	19.26 ± 0.02	24.32 ± 0.02	19.95 ± 0.01	4.73 ± 0.01	4.48 ± 0.02	5.09 ± 0.01
Stn. 11	20.50 ± 0.01	25.93 ± 0.02	20.90 ± 0.02	5.13 ± 0.02	4.90 ± 0.01	5.23 ± 0.02
Stn. 12	22.97 ± 0.02	29.79 ± 0.02	24.55 ± 0.01	5.11 ± 0.02	4.78 ± 0.03	5.34 ± 0.03

Phase A = pre-Yash period (17.05.2021), Phase B = Yash phase (26.05.2021), Phase C = post-Yash phase (08.06.2021); Units of surface water salinity and DO are ppt and ppm respectively

It is interesting to note that during all these three super cyclones, the aquatic salinity increased considerably during the cyclonic phase (Mitra et al. 2011, 2020) that resulted to massive breaching of the embankments leading to flooding of the agricultural lands and freshwater ponds in the island villages (Figs. 2.3, 2.4 and 2.5).

The dissolved oxygen (DO) decresed simultaneously in the selected stations due to intrusion of saline water from the offshore region as pointed out by several researchers (Mitra et al. 2009) in this region (Figs. 2.6, 2.7 and 2.8).

The decrease of dissolved oxygen due to increase of salinity posed negative impact on the fish community of the study region and tonnes of fishes were found dead after the super cyclones causing a dip in the sector of fishery (Mitra et al. 2011).

Apart from this, adverse impact on deep sea fishing was also observed which is one of the major livelihoods of the island dwellers in Sundarban delta.

The intrusion of saline water also damaged the agricultural crops preferably the paddy as this major crop cannot withstand very moderate or high salinity. We observed that in most of the paddy field the salinity exceeded 25 psu (Fig. 2.9).

Six key components usually regulate the livelihood profile in a region. These are highlighted here with respect to the framework of Indian Sundarbans.

1. Biophysical component—This component focuses on loss of access to natural resources, environmental change, and constraints, which in case of Indian Sundarbans encompasses alteration of salinity, frequent natural disasters, pollution or deterioration of water quality, acidification (Mitra and Zaman 2021), over exploitation etc.

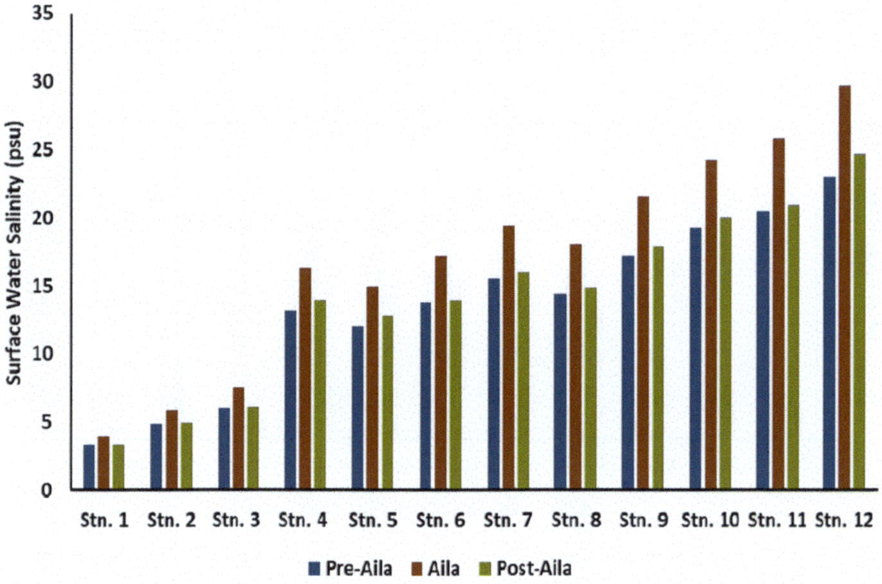

Fig. 2.3 Alteration of aquatic salinity (in psu) due to super cyclone Aila in 2009 in and around 12 selected stations in Indian Sundarbans

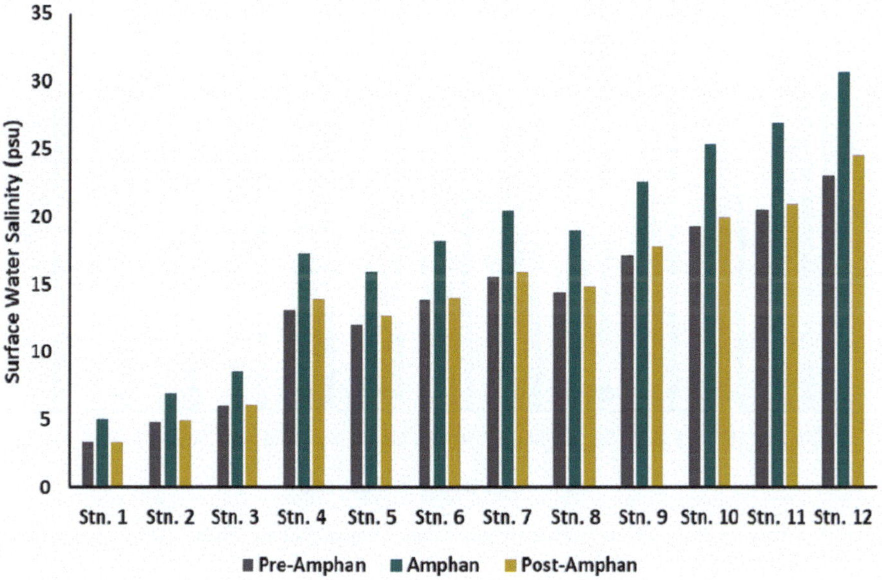

Fig. 2.4 Alteration of aquatic salinity (in psu) due to super cyclone Amphan in 2020 in and around 12 selected stations in Indian Sundarbans

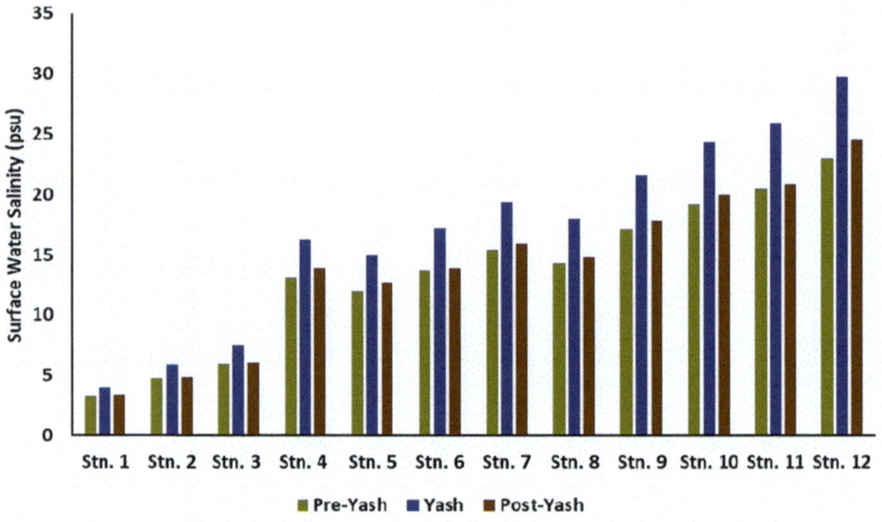

Fig. 2.5 Alteration of aquatic salinity (in psu) due to super cyclone Yash in 2021 in and around 12 selected stations in Indian Sundarbans

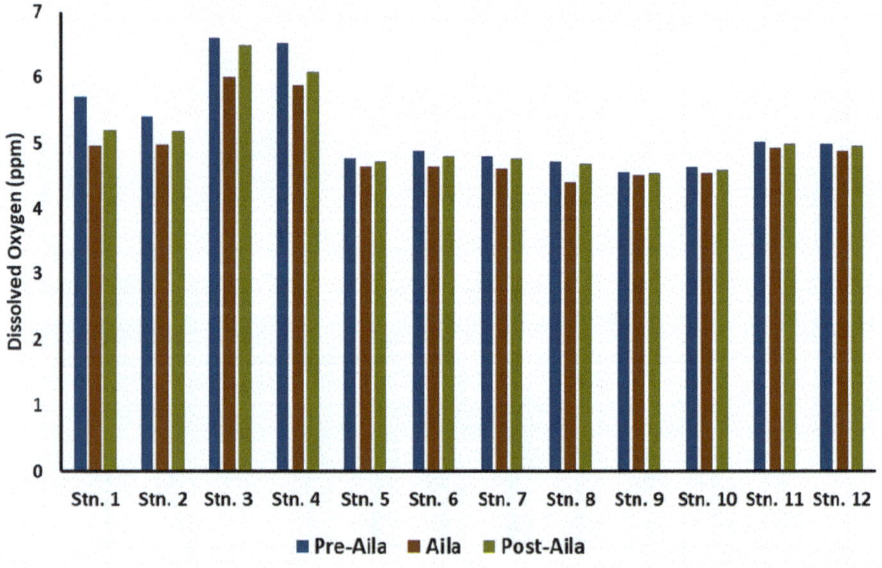

Fig. 2.6 Decrease of dissolved oxygen (in ppm) due to super cyclone Aila in 2009 in and around 12 selected stations in Indian Sundarbans

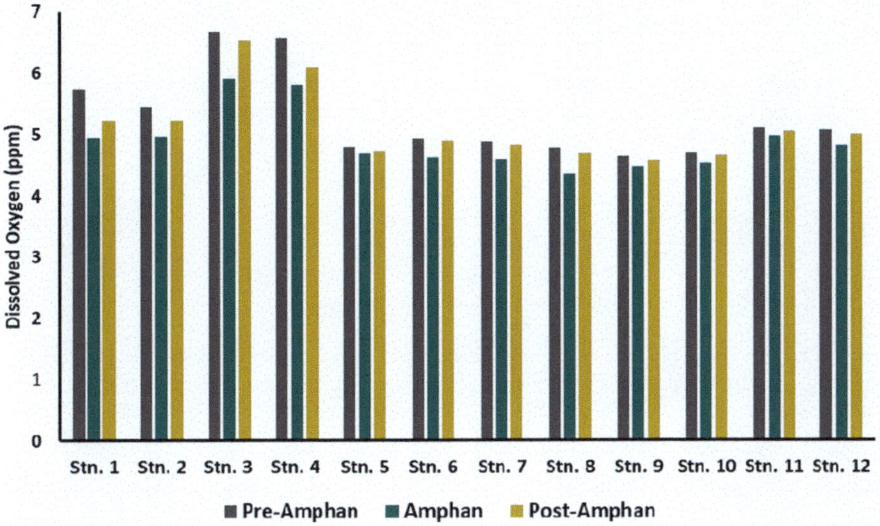

Fig. 2.7 Decrease of dissolved oxygen (in ppm) due to super cyclone Amphan in 2020 in and around 12 selected stations in Indian Sundarbans

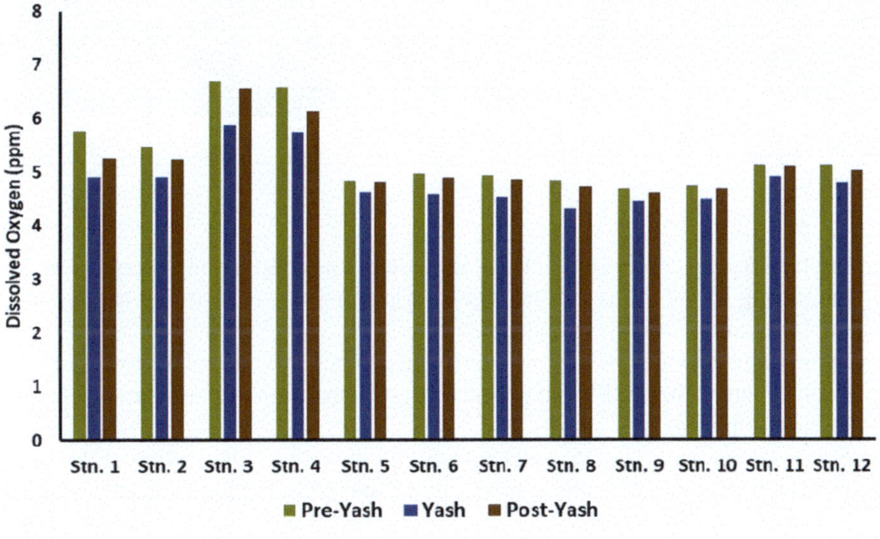

Fig. 2.8 Decrease of dissolved oxygen (in ppm) due to super cyclone Yash in 2021 in and around 12 selected stations in Indian Sundarbans

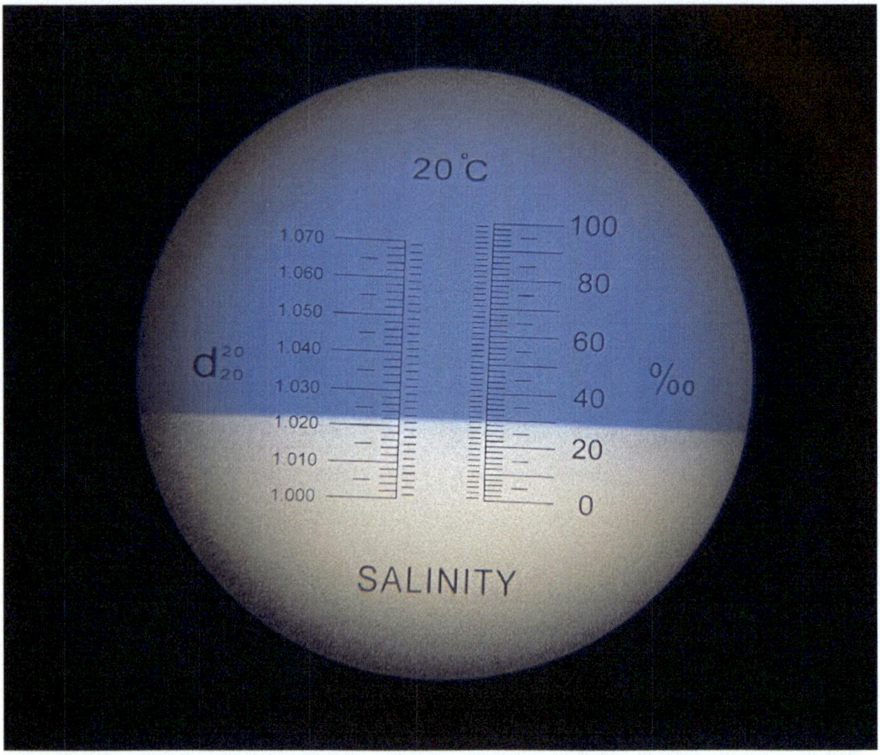

Fig. 2.9 Aquatic salinity touching almost 30 psu was observed in refractometer after the super cyclone Yash in the paddy field of Indian Sundarbans

2. Political/legal component—This includes rules/regulations connected to harnessing of natural resources, decision making power, institutional support, implementation of new/amended laws etc.
3. Economic component—It is one of the vital components that encompasses opportunities, innovation, skill, and productivity. Competition is also a part and parcel of the component.
4. Social component—Social values primarily depend on the historical development of the region, education, productivity, and conservation measures adopted to protect the natural resources. The eastern sector of Indian Sundarbans under the Reserve Forest (RF) area is an important example in this context where the natural resources are conserved due to strict vigilance of the West Bengal Forest Department. This accelerates the productivity that can scale up the livelihood opportunities.
5. Cultural component—It includes the traditional practices and beliefs as seen in Indian Sundarbans. In this deltaic complex, Bonobibi is worshipped by people of all religions with the belief that they will be protected from wild animals

(preferably tiger) while procuring natural resources like timber, fishes, crabs, honey, wax, fodder etc. from the deep mangrove forest.

6. Psychological component—It includes stress and outlook, motivation/mind set for participation in a new project/alternative livelihoods, uncertainty to initiate a new type of livelihood (like oyster culture, seaweed culture or halophyte farming) in Indian Sundarbans.

The island dwellers of Sundarbans are mostly scared to take new ventures as there is uncertainty of markets for oysters, seaweeds, or any mangrove-based snacks that was carried in pilot scale sporadically in few pockets of Indian Sundarbans by many researchers (Mitra et al. 2000a, b; Mukherjee et al. 2007; Bhattacharyya et al. 2010; Ghosh et al. 2011; Pramanick et al. 2015, 2016, 2017; Biswas et al. 2019; Mitra et al. 2019; Biswas et al. 2020; Pramanick et al. 2020, 2021; Yadav and Majumdar 2020).

2.1 Agriculture

Agriculture is one of the main livelihoods for the island dwellers of Indian Sundarbans. The major crop cultivated in the islands of Indian Sundarbans is Aman (Kharif). The production of this crop is a function of monsoonal pattern in the area, which has become highly irregular in the lower Gangetic delta complex (Table 2.4).

Alteration in monsoon pattern and inundation of agricultural field with saline water causes poor production of crop. A pilot scale initiative was taken in 2017 for introduction of salt tolerant paddy variety in place of high-yield variety, but the success rate was extremely poor. People in Indian Sundarbans grow vegetables, pulses, oilseeds, chilies, tomatoes etc. during the premonsoon and postmonsoon seasons. Betel leaf is cultivated in many areas of western Indian Sundarbans.

In the framework of Indian Sundarbans, the agricultural scenario comes under CDR (Complex, Diverse and Risk prone) system. The main reason behind the risk system is the alteration of salinity, which is rising in the central sector of Indian Sundarbans (Table 2.5 and Fig. 2.10), but decreasing in the western part (Table 2.6 and Fig. 2.11). Similar observations were also recorded by earlier researchers in the present study area (Banerjee et al. 2010, 2013; Chand et al. 2012; Chowdhury et al. 2021; Dhar 2011; Dhar et al. 2021; Guha and Mitra 2020; Mitra 2013, 2018; Mitra and Zaman 2015, 2016, 2020, 2021; Mitra et al. 2010; Pal et al. 2017; Trivedi et al. 2016).

Another important risk in connection to agriculture is the lowering of aquatic pH both in the western (Table 2.7 and Fig. 2.12) and central (Table 2.8 and Fig. 2.13) sectors that may be attributed to steady rise of carbon dioxide in Indian Sundarbans due to industrial development (in the adjoining area like Haldia), deforestation, unplanned urban settlement and flourishing of shrimp farms and unregulated tourism units at the cost of mangroves (Mira and Zaman 2021).

Table 2.4 Temporal variation of rainfall pattern

Year	Jan	Feb	Mar	Apr	May	Jun	Jul	Aug	Sep	Oct	Nov	Dec
1981	–	–	–	–	–	–	–	–	–	–	0	0.38
1982	0	1.69	1.65	2.77	0.85	5.69	8.49	10.4	4.23	0.36	0.87	0
1983	0.14	1.67	1.2	1.08	2.51	11.22	6.42	17.79	9.96	3.61	0.25	0.18
1984	0.57	0.03	0	2.29	3.19	23.81	7.26	15.55	6.69	3.28	0	0
1985	0.88	0.21	0.2	1.12	5.11	9.25	9.25	12.2	7.69	3.86	0	0
1986	0.33	0	0.05	1.04	6.2	6.34	10.66	3.35	27.91	8.55	7.54	0.05
1987	0	0.7	0.6	5.8	4.21	4.73	11.55	13.41	9.58	0.81	2	0.32
1988	0	0.92	0.12	0.71	5.49	20.29	17.4	9.28	6.38	4.89	4.66	0
1989	0.07	0.25	0.32	0	5.15	2.83	9.1	10.35	11.34	6.63	0.39	0.08
1990	0	2.48	6.35	3.27	4.79	10.18	16.6	10.77	10.4	5.72	2.41	0.06
1991	2	0.64	0.55	1.13	0.49	16.21	8.64	8.07	8.02	4.56	0.02	0.57
1992	0.92	1.73	0	0.59	6.1	13.31	12.13	13.51	16.1	3.8	0	0
1993	0.03	0	2.75	0.77	4.61	10.34	11.34	11.55	17.73	3.97	0.43	0
1994	0.11	0.84	0.25	3.82	2.35	8.21	14.93	9.12	6.46	2.52	1.92	0
1995	0.23	1.36	0.08	0.61	7.39	11.37	17.6	10	17.84	3.08	6.68	0
1996	0.29	0.3	0.07	0.89	2.56	18.97	11.88	16.72	5.09	9.94	0	0
1997	0.57	0.49	1.36	6.64	3.41	7.12	12.98	15.78	9.36	0.16	0.49	0.48
1998	0.98	0.14	5.97	2.13	5.86	4.94	12.07	12.09	10.41	5.31	5.84	0
1999	0.08	0	0.01	0.05	6.72	8.94	16.47	13.38	14.91	5.96	0.07	0
2000	0.05	1.96	0	2.29	10.79	7.79	15.21	6.34	11.69	4.22	0.06	0.06
2001	0.09	0.01	0.8	3.43	6.8	14.54	10.63	6.87	5.84	5.14	0.62	0
2002	0.85	0	0.53	2.66	4.01	19.81	8.78	10.05	8.31	3.97	4.62	0

(continued)

Table 2.4 (continued)

Year	Jan	Feb	Mar	Apr	May	Jun	Jul	Aug	Sep	Oct	Nov	Dec
2003	0	0.51	0.9	2.26	4.36	10.65	11.9	11.49	9.37	17.09	0.67	1.29
2004	0.15	0.25	0.13	2.26	2.42	8.96	13.35	14.93	7.75	8.77	0	0
2005	1.47	0	4.25	1.24	2.27	8.3	17.89	9.17	11.73	16.74	0	0.03
2006	0	0	0.11	2.06	3.9	4.08	21.16	14.55	17.35	1.045	0.03	0
2007	–	2.8	0.07	1.74	5.63	5.32	20.25	8.97	21.39	3.9	2.13	0
2008	3.46	0.34	0.15	1.4	2.26	12.22	7.16	8.96	9.62	6	–	–
2009	0	0	0.53	0.03	6.52	1.79	13.19	11.47	10.05	4.93	0.34	0
2010	0	0.13	0	0.72	4.25	8.44	8.53	7.56	7.24	4.51	0.12	0.36
2011	0	0.24	1.2	7.31	3.34	14.09	8.98	22.56	7.86	3.1	–	0
2012	1.57	1.01	0.17	3.15	1.52	6.08	10.2	14.78	8.35	4.28	1.86	1.15
2013	0.12	0.18	0.15	0.63	6.36	8.25	8.65	22.08	14.84	9.3	0	0
2014	0	2.33	1.77	0	–	5.78	7.35	–	11.37	1.7	0	0
2015	0.14	0.19	1.83	0.74	4.24	8.21	6.80	13.14	20.14	2.85	0	0
2016	0	0.07	1.09	0	5.10	4.98	5.93	9.68	12.65	3.10	0.12	0
2017	0	0.14	0	0	1.26	11.02	7.80	12.65	9.89	4.02	0	0.14
2018	0.10	1.02	1.21	0.35	3.58	13.65	11.66	14.20	8.42	5.17	0	1.02
2019	1.23	1.45	1.66	0.41	2.97	17.04	13.45	12.06	9.17	8.03	1.05	1.33
2020	1.42	1.27	1.54	0.66	1.56	13.65	12.04	9.28	10.44	11.55	2.40	1.91
2021	1.18	1.34	0.22	0.38	1.11	15.39	19.56	13.76	12.15	14.59	3.26	2.54

Source IMD (Indian Meteorological Department); '–' means data not available; Jan—January; Feb—February; Mar—March; Apr—April; May—May; Jun—June; Jul—July; Aug—August; Sep—September; Oct—October; Nov—November; Dec—December

Table 2.5 Seasonal variation of surface water salinity (psu) in Bali Island (22°04′35.17″N and 88°44′55.70″E) located in central Indian Sundarbans

Year	Premonsoon	Monsoon	Postmonsoon
1984	22.10	20.70	21.84
1985	23.62	20.77	22.11
1986	24.37	20.83	22.05
1987	24.25	20.17	22.70
1988	25.07	20.06	22.43
1989	25.72	20.38	23.11
1990	25.87	21.05	23.57
1991	25.04	21.72	23.60
1992	24.75	21.76	23.99
1993	25.63	21.89	23.80
1994	26.03	21.77	24.24
1995	26.07	22.02	25.02
1996	26.11	21.64	25.08
1997	26.86	22.04	24.95
1998	26.84	22.09	25.46
1999	26.78	20.67	25.44
2000	27.01	21.40	25.54
2001	27.07	22.63	25.99
2002	27.16	22.95	26.45
2003	27.48	23.01	26.51
2004	27.23	23.07	26.48
2005	27.21	23.19	27.59
2006	27.89	23.81	27.41
2007	27.72	23.86	27.72
2008	28.37	23.95	27.75
2009	31.71	22.90	28.33
2010	29.90	22.99	28.57
2011	30.16	23.4	28.92
2012	30.11	23.55	28.75
2013	30.49	23.72	28.95
2014	30.76	24.07	29.31
2015	30.84	23.92	29.83
2016	31.12	24.06	29.91
2017	30.83	25.06	29.33
2018	30.99	25.44	30.71
2019	30.97	25.76	30.68

(continued)

Table 2.5 (continued)

Year	Premonsoon	Monsoon	Postmonsoon
2020	31.02	25.82	30.83
2021	31.05	26.02	30.89
2022	31.63	26.14	30.94

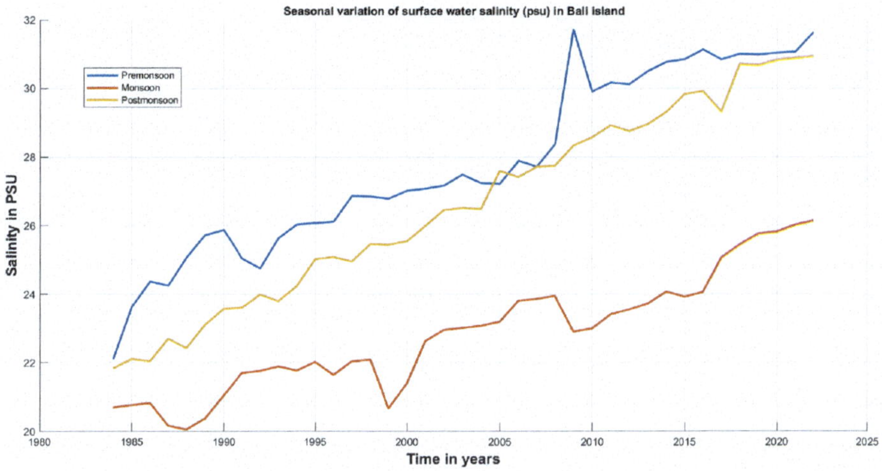

Fig. 2.10 Increasing trend of aquatic salinity around Bali Island (22°04′35.17″N and 88°44′55.70″E) in central Indian Sundarbans

Table 2.6 Seasonal variation of surface water salinity (psu) in Kakdwip (21°52′26.50″N and 88°08′04.48″E) located in western Indian Sundarbans

Year	Premonsoon	Monsoon	Postmonsoon
1984	17.53	12.99	15.02
1985	16.89	12.95	14.96
1986	16.72	12.82	14.98
1987	16.84	11.85	14.02
1988	17.4	11.69	14.66
1989	16.87	11.41	14.18
1990	16.7	11.87	13.52
1991	16.17	10.98	12.99
1992	15.92	10.92	13.22
1993	15.82	10.84	13.81
1994	15.3	10.69	12.68

(continued)

Table 2.6 (continued)

Year	Premonsoon	Monsoon	Postmonsoon
1995	15.48	10.89	12.25
1996	14.71	10.82	12.29
1997	14.16	10.51	12.52
1998	13.55	10.29	11.97
1999	13.8	9.81	11.99
2000	12.89	9.63	12.04
2001	13.28	9.27	11.28
2002	12.36	9.14	11.26
2003	12.1	9.4	11.62
2004	12.05	8.84	11.69
2005	12.24	8.32	11.08
2006	11.62	8.61	11.20
2007	11.02	8.19	11.28
2008	10.92	8.84	10.25
2009	16.89	9.20	10.05
2010	10.85	8.08	10.02
2011	10.51	7.91	9.85
2012	10.86	7.84	9.94
2013	10.78	7.8	9.78
2014	10.11	7.77	9.71
2015	10.92	7.72	9.56
2016	9.94	6.93	9.72
2017	9.97	6.86	9.40
2018	8.91	6.7	8.02
2019	8.84	6.63	7.94
2020	8.55	5.44	7.06
2021	8.01	4.79	6.88
2022	7.22	3.13	6.31

In addition to the above risks, water quality deterioration in and around the agricultural fields has posed severe negative impact on the quality of crops preferably paddy.

There is significant discrepancy in the statistical figures of cultivable land in Indian Sundarbans. The area ranges between 3.06 lakh ha to 4.71 lakh ha out of which 65,000 ha is used for double cropping. The agricultural land in Indian Sundarbans can be divided into three types, based on the elevation (Table 2.9).

Apart from paddy and different types of vegetables, cultivation of betel vine fetches considerable economic return to the people of Indian Sundarbans. The betel

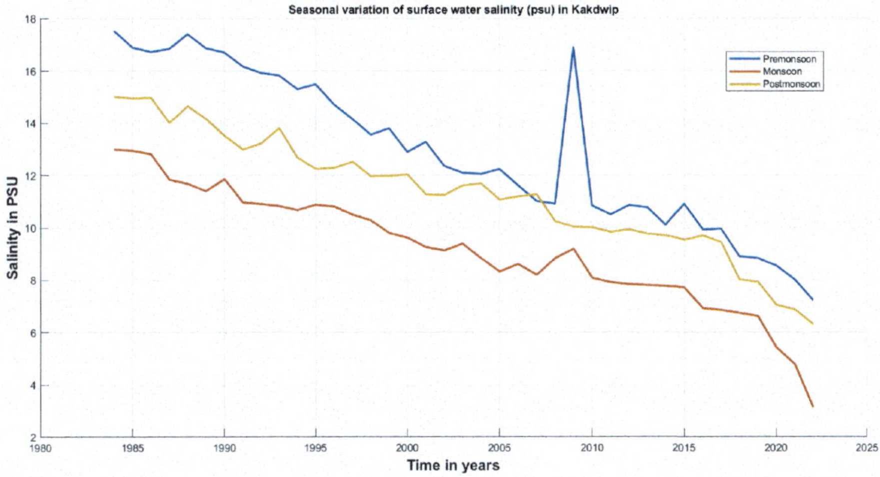

Fig. 2.11 Decreasing trend of aquatic salinity around Kakdwip (21°52′26.50"N and 88°08′04.48"E) in western Indian Sundarbans

Table 2.7 Seasonal variation of surface water pH in Chemaguri (21°39′42.88″N and 88°08′49.01″E) located in western Indian Sundarbans

Year	Premonsoon	Monsoon	Postmonsoon
1984	8.31	8.29	8.3
1985	8.31	8.3	8.31
1986	8.31	8.3	8.3
1987	8.3	8.28	8.3
1988	8.31	8.29	8.31
1989	8.3	8.3	8.31
1990	8.3	8.28	8.3
1991	8.3	8.29	8.3
1992	8.3	8.28	8.3
1993	8.31	8.29	8.3
1994	8.3	8.29	8.3
1995	8.3	8.28	8.3
1996	8.31	8.3	8.29
1997	8.3	8.29	8.3
1998	8.3	8.29	8.3
1999	8.29	8.28	8.28
2000	8.29	8.28	8.29
2001	8.29	8.28	8.29

(continued)

Table 2.7 (continued)

Year	Premonsoon	Monsoon	Postmonsoon
2002	8.3	8.28	8.28
2003	8.29	8.28	8.29
2004	8.29	8.27	8.28
2005	8.3	8.28	8.3
2006	8.29	8.28	8.29
2007	8.28	8.27	8.28
2008	8.28	8.26	8.28
2009	8.32	8.27	8.27
2010	8.28	8.27	8.28
2011	8.28	8.26	8.28
2012	8.26	8.25	8.26
2013	8.25	8.24	8.25
2014	8.25	8.23	8.25
2015	8.26	8.23	8.24
2016	8.25	8.21	8.24
2017	8.23	8.19	8.22
2018	8.21	8.15	8.19
2019	8.20	8.18	8.19
2020	8.21	8.17	8.18
2021	8.21	8.16	8.17
2022	8.2	8.15	8.17

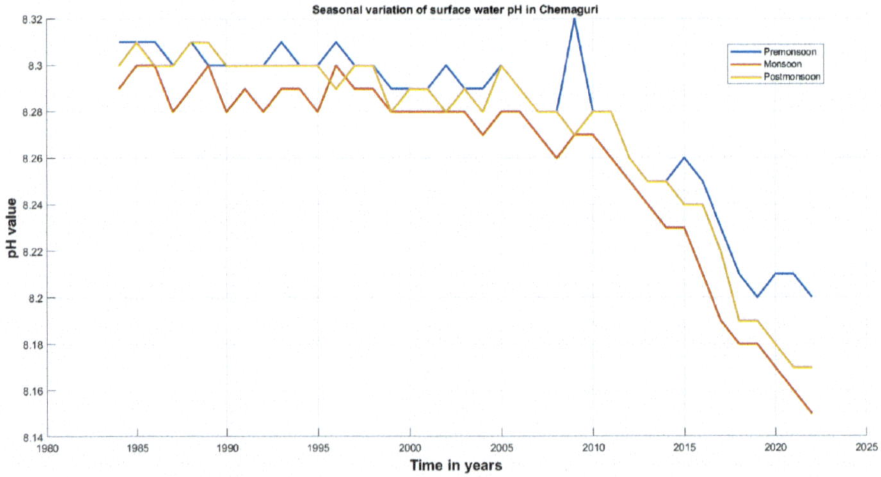

Fig. 2.12 Decreasing trend of aquatic pH around Chemaguri (21°39′42.88″N and 88°08′49.01″E) in western Indian Sundarbans

Table 2.8 Seasonal variation of surface water pH in Chotomollakhali (22°10′40.00″N and 88°54′26.71″E) located in central Indian Sundarbans

Year	Premonsoon	Monsoon	Postmonsoon
1984	8.33	8.31	8.32
1985	8.33	8.31	8.32
1986	8.33	8.31	8.32
1987	8.32	8.3	8.31
1988	8.31	8.31	8.31
1989	8.31	8.31	8.32
1990	8.32	8.3	8.32
1991	8.32	8.3	8.32
1992	8.32	8.31	8.32
1993	8.32	8.31	8.31
1994	8.32	8.3	8.32
1995	8.3	8.31	8.32
1996	8.3	8.29	8.31
1997	8.29	8.29	8.31
1998	8.3	8.3	8.31
1999	8.29	8.29	8.3
2000	8.3	8.29	8.3
2001	8.3	8.29	8.3
2002	8.31	8.29	8.31
2003	8.31	8.29	8.3
2004	8.29	8.29	8.29
2005	8.29	8.29	8.29
2006	8.28	8.29	8.28
2007	8.28	8.3	8.29
2008	8.29	8.28	8.29
2009	8.33	8.29	8.29
2010	8.3	8.28	8.28
2011	8.3	8.27	8.28
2012	8.3	8.26	8.27
2013	8.28	8.26	8.27
2014	8.28	8.25	8.27
2015	8.27	8.24	8.26
2016	8.26	8.23	8.24
2017	8.25	8.2	8.24
2018	8.25	8.18	8.21
2019	8.24	8.18	8.22
2020	8.23	8.16	8.20
2021	8.22	8.15	8.19
2022	8.22	8.15	8.19

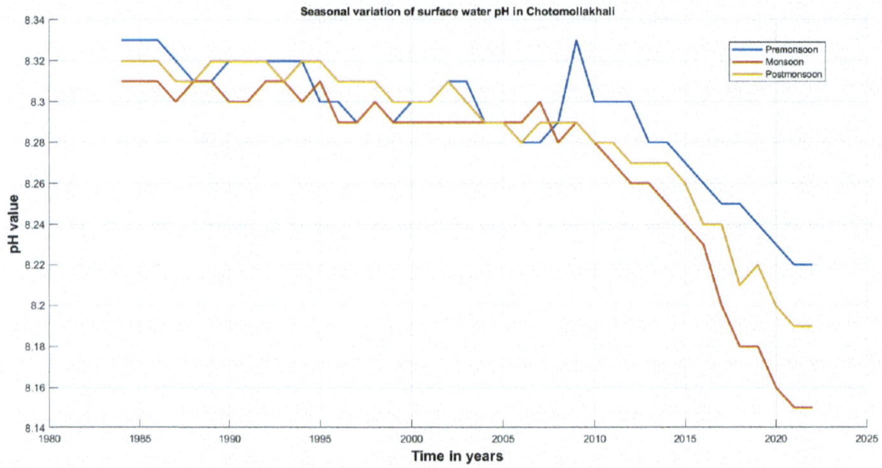

Fig. 2.13 Decreasing trend of aquatic pH around Chotomollakhali (22°10′40.00″N and 88°54′26.71″E) in central Indian Sundarbans

Table 2.9 Type for land used for agriculture in Indian Sundarbans

Land type	%	Features	Crop variety
High land	11	Water table above the ground is around 15–20 cm	Paddy and winter vegetables
Medium upland	26	Water table above the ground is between 20–30 cm	Paddy and vegetables
Low land	61	Water table in this category of land is around 50 cm	Aman paddy

leaves locally known as 'Paan' has great demand in several Asian countries like Thailand, Malaysia, Indonesia, Myanmar, Pakistan, and Bangladesh. Betel vines are grown in closed bamboo made structures, which are locally known as 'Baroj'. These are built to protect the plant from the scorchy heat of the sun in premonsoon and cold wind in postmonsoon. The structures are rectangular in shape and normally range between 30–40 sq. m (Fig. 2.14). We carried out a study during March, 2023 on the socio-economic status of the betel vine farmers in Sagar Island, the largest island in Indian Sundarbans (Table 2.10) and observed that a large percentage of the agriculturists are dependent on betel vine cultivation.

Fig. 2.14 Rectangular structure of betel leaves (locally known as Paan Baroj)

Table 2.10 Socio-economic profile of the betel vine farmers in study area (n = 86 families)

Sl. No	Particulars	Value
A	Details of the family at Sagar Island engaged in betel vine cultivation	
	(i) Family Size (No.) considering mean of 65 families	6.12
B	Livelihood	
	(i) Agriculture (%)	77.32
	(ii) Pisciculture (%)	15.83
	(iii) Prawn seeds collectors (%)	4.92
	(iv) Daily labour (mainly engaged in boat repairing & fishing vessels conditioning) (%)	1.93
C	Age of the betel vine cultivators	
	(i) Age 20–40 years (%)	59.40
	(ii) Age 41–55 years (%)	38.24
	(iii) Age 55–70 years (%)	2.36
D	Operational holding size (mean of 27 Baroj) (ha)	0.55
E	Materials required for betel vine cultivation	
	(i) Own material (%)	15
	(ii) Materials obtained from different sources (%)	85
F	Family income in INR (per year) (mean of 48 families)	75,800

2.2 Pisciculture and Fishing

The fishery sector occupies an important sector in the livelihood domain of Indian sub-continent. The inland and marine fish landings fetch considerable revenue through which livelihoods in several tiers of Indian society are sustained. Data on inland and marine fish landings by the Indian states and Union Territories are provided in Tables 2.11, 2.12 and 2.13 which reflect the species-wise catch volume.

It is to be noted in this context that although the record of fish consumption data during 2019–20 is not available for West Bengal (Table 2.14, *vide* row No. 28), but our ground level observation shows the quantum of fish landing in and around Indian Sundarban region, which is quite high in volume specially the brackish water fishes (Table 2.15).

Considering the congenial environment of aquatic ecosystem in terms of fish production, people of Sundarbans has initiated monoculture of prawns, polyculture of prawns with other species etc. This, however, has resulted mass destruction of mangroves as most of the farms are being operated within the heart of the halophytic forest (Fig. 2.15).

Threat being the other side of opportunity also prevails in Sundarbans that damages the fishery and aquaculture sectors to a great extent almost in every year. The threat is keenly related to the natural disasters preferably cyclone that hit the lower Gangetic region almost every year causing a bulk intrusion of saline water from the Bay of Bengal region to the water bodies in and around the islands of Sundarbans. Due to high salinity, the dissolved oxygen reduces causing considerable stress on the fish species both in the wild as well as the culture system. The alterations of salinity and dissolved oxygen in twelve selected stations (*vide* Fig. 2.2 and Tables 2.1, 2.2 and 2.3) are highlighted in Figs. 2.16, 2.17, 2.18, 2.19, 2.20, 2.21, 2.22, 2.23, 2.24, 2.25, 2.26, 2.27, 2.28, 2.29, 2.30, 2.31, 2.32, 2.33.

2.3 Animal Husbandry

People of Indian Sundarbans also earn money by rearing cow, goat, sheep, and pig (Figs. 2.34 and 2.35).

In recent times, white pigs are reared for pork, which provides a stable income to the people of Indian Sundarbans (Fig. 2.36). The rearing of this species has been given thrust mainly because of their growth in biomass and high rate of breeding frequency; pigs can breed frequently and give birth to around eight to ten piglets during each cycle.

The livestock sector contributes about 22% of the total value of output in agriculture, fishing, and forestry sectors. The lower Gangetic delta is blessed with the presence of a valuable genetically rich sheep and goat known as "Garole Sheep" and "Black Bengal Goat" respectively. The livestock sector of Indian Sundarbans produces several marketable items like meat, milk, fiber, skin etc. We carried out the

Table 2.11 Species-wise inland fish landings by states and union territories: 2019–20 (in lakh tonnes)

States/UT's	Major carps (Catla, Rohu, Mrigal)	Minor carps	Exotic carps	Murrels	Catfishes	Other fresh water fishes	Total
Andhra Pradesh	19.98	0	2.19	0.58	2.1	11.26	36.11
Arunachal Pradesh	0.01	0	0.02	0	0	0.02	0.05
Assam	1.44	0.48	0.27	0.25	0.4	0.89	3.73
Bihar	3.14	1.41	0.13	0.45	0.06	1.22	6.41
Chhattisgarh	4.32	0.05	0.52	0.01	0.43	0.39	5.72
Goa	0	0	0	0	0	0.04	0.04
Gujarat	0.41	0.01	0	0.02	0.06	1.08	1.58
Haryana	1.17	0.02	0.66	0	0.03	0.02	1.9
Himachal Pradesh	0.13	0	0	0	0	0.01	0.14
Jharkhand	1.83	0.02	0.2	0	0.18	0	2.23
Karnataka	1.04	0.17	0.78	0	0.02	0.28	2.29
Kerala	0.1	0	0.01	0	0.02	1.92	2.05
Madhya Pradesh	1.54	0.1	0.29	0	0.06	0.01	2
Maharashtra	0.78	0	0.12	0	0	0.28	1.18
Manipur	0.1	0.01	0.2	0	0	0.01	0.32
Meghalaya	0.06	0.02	0.05	0	0	0.01	0.14
Mizoram	0.03	0	0.04	0	0	0	0.07
Nagaland	0.05	0	0.04	0	0	0	0.09
Odisha	3.96	0.4	0.79	0.04	0.11	1.3	6.6
Punjab	0.85	0.02	0.61	0	0.02	0.01	1.51
Rajasthan	0.42	0.1	0.04	0	0.02	0.58	1.16
Sikkim	0	0	0	0	0	0	0

(continued)

Table 2.11 (continued)

States/UT's	Major carps (Catla, Rohu, Mrigal)	Minor carps	Exotic carps	Murrels	Catfishes	Other fresh water fishes	Total
Tamil Nadu	0.54	0	0.21	0.03	0.05	0.91	1.74
Telangana	1.45	0.16	0.4	0.38	0.2	0.41	3
Tripura	0.51	0.08	0.17	0	0.01	0.01	0.78
Uttar Pradesh*	5.61	1.27	–	0	0.08	0.03	6.99
Uttarakhand	0.02	0	0.01	0	0.01	0.01	0.05
West Bengal	9.96	0.55	1.93	0.25	0.44	3.06	16.19
Andaman and Nicobar Islands	0	0	0	0	0	0	0
Chandigarh	0	0	0	0	0	0.01	0.01
Daman and Diu, Dadra & Nagar haveli	0	0	0	0	0	0	0
Delhi	0.01	0	0	0	0	0	0.01
Jammu & Kashmir	0.01	0.01	0.12	0	0.01	0.06	0.21
Ladakh	0	0	0	0	0	0	0
Lakshadweep	0	0	0	0	0	0	0
Puducherry	0.03	0.01	0.01	0	0.01	0.01	0.07
Total	59.5	4.89	9.81	2.01	4.32	23.84	104.37

Source Department of Fisheries, States Government/UTs Administration
* Uttar Pradesh: 561,000 (Including Exotic Carps); https://dof.gov.in/sites/default/files/2021-02/Final_Book.pdf

Table 2.12 Species-wise marine fish landings by coastal states and union territories: 2019–20 (in lakh tonnes) (East Coast)

Species	Andhra Pradesh	Odisha	Tamil Nadu	West Bengal	A and N Islands	Puducherry	Total
Sharks (Selachimorpha)	0.18	0.02	0	0	0	0	0.2
Skates/Guitarfish (Rajidae)	0.14	0.01	0.04	0.09	0.02	0	0.3
Rays (Batoidea)	0.09	0.01	0	0	0	0.01	0.11
Eels (Anguilliformes)	0.08	0.03	0.08	0	0	0	0.19
Catfishes (Siluriformes)	0.21	0.12	0.01	0.2	0	0.01	0.55
Indian Oil sardine (Sardinella longiceps)	0.3	0.07	0.06	0.05	0.04	0.05	0.57
Other sardines (*Sardinella fimbriata* and others)	0.17	0.04	0.3	0	0	0.03	0.54
Hilsa shad (*Tenualosa ilisha*)	0.08	0.04	0.44	0.2	0	0	0.76
Other shads (Alosinae)	0.03	0.04	0	0	0	0	0.07
Coilia (*Coilia* spp.)	0.03	0	0	0	0	0	0.03
Setipinna (*Setipinna* spp.)	0.03	0	0	0	0	0	0.03
Stolephorus (*Stolephorus* spp.)	0.08	0	0	0	0	0	0.08
Thryssa (*Thryssa* spp.)	0.05	0.03	0.23	0	0	0	0.31
Wolf herring (*Chirocentrus* spp.)	0.01	0.02	0.04	0	0	0	0.07
Other clupeids	0.45	0.1	0.26	0	0	0.01	0.82
Bombay-duck (*Harpadon nehereus*)	0.04	0.02	0	0	0	0	0.06
Lizard fishes (*Synodontidae*)	0.03	0	0.06	0	0	0.01	0.1
Half beaks & full beaks (*Hemiramphidae & Belonidae*)	0.01	0	0.06	0	0	0	0.07
Flying fishes (*Exocoetidae*)	0.01	0	0.11	0	0	0.01	0.13
Rock cods (*Lotella rhacina*)	0.03	0	0	0	0	0	0.03
Snappers (*Lutjanidae*)	0.01	0.01	0	0	0.05	0	0.07
Pig-face bream (*Lethrinidae*)	0.03	0	0.14	0	0	0	0.17

(continued)

Table 2.12 (continued)

Species	Andhra Pradesh	Odisha	Tamil Nadu	West Bengal	A and N Islands	Puducherry	Total
Threadfin bream (*Nemipteridae*)	0.09	0.02	0.09	0	0	0.01	0.21
Grouper (*Epinephelinae*)	0.09	0	0	0	0	0	0.09
Cobia (*Rachycentron canadum*)	0	0	0.01	0	0	0	0.01
Seabass (*Lates calcarifer*)	0	0	0	0	0	0.01	0.01
Other perches & perch like	0.07	0.02	0.26	0.09	0	0.01	0.45
Goatfishes (Mullidae)	0.09	0	0.15	0	0	0	0.24
Threadfins (Polynemidae)	0.08	0.03	0.05	0	0	0.01	0.17
Croakers (Sciaenidae)	0.08	0.05	0.15	0.12	0.01	0	0.41
Ribbon fishes (Trachipteridae)	0.12	0.05	0.09	0	0	0.01	0.27
Horse Mackerel (*Caranx & Trichiurus*)	0.01	0.01	0.25	0.2	0	0.01	0.48
Scads (*Decapterus* spp.)	0.01	0	0	0	0	0	0.01
Leather-jackets (*Oligoplites saurus*)	0.01	0	0	0	0	0	0.01
Pompano (*Trachynotus* spp.)	0	0.01	0.02	0	0	0	0.03
Other carangids	0	0	0	0.1	0.03	0	0.13
Silverbellies (Gerridae)	0.01	0.01	0.3	0	0.03	0.02	0.37
Big-jawed jumper (*Lactarius lactarius*)	0	0	0.07	0	0	0	0.07
Black pomfret (*Parastromatius niger*)	0.28	0.04	0.01	0	0	0.01	0.34
Silver pomfret (*Pampus argenteus*)	0.15	0.03	0.01	0	0	0	0.19
Indian mackerel (*Rastrelliger kanagurta*)	0.19	0.06	0.24	0.08	0	0.02	0.59
Other mackerels	0.13	0.01	0	0	0.03	0	0.17
Scomberomorus spp.	0	0.04	0.1	0	0.01	0.01	0.16
Acanthocybium solandri	0	0	0	0	0	0	0
Kawa kawa (*Euthynnus affinis*)	0	0.01	0.26	0	0.02	0	0.29
Frigate tuna (*Auxis* spp.)	0.01	0	0	0	0	0	0.01

(continued)

Table 2.12 (continued)

Species	Andhra Pradesh	Odisha	Tamil Nadu	West Bengal	A and N Islands	Puducherry	Total
Skipjack tuna (*Katsuwonus pelamis*)	0.03	0	0	0	0	0	0.03
Longtail tuna (*Thunnus tonggol*)	0.01	0	0	0	0	0.01	0.02
Yellowfin tuna (*Thunnus albacares*)	0.12	0	0	0	0	0	0.12
Other tunnies	0.11	0	0	0	0	0	0.11
Bill fishes (Xiphiidae)	0	0	0	0	0	0	0
Barracudas (*Sphyraena* spp.)	–	0	0.08	0	0.01	0.01	0.1
Mullets (Mugilidae)	0.12	0.06	0.03	0.1	0.01	0.04	0.36
Unicorn cod (*Bregmaceros* mcclellandii)	–	0	0	0	0	0	0
Halibut (*Hippoglossus* spp.)	0.04	0	0	0	0	0	0.04
Flounders (Pleuronectoidei)	–	0	0	0	0	0	0
Soles (Soleidae)	–	0.02	0.07	0	0	0.01	0.1
Penaeid prawns (Penaeidae)	1.08	0.4	0.02	0.07	0	0.03	1.6
Non-penaeid prawns (Sergestidae)	0.18	0.04	0.38	0.03	0	0	0.63
Lobsters (Nephropidae)	0.01	0	0.03	0	0	0	0.04
Crabs (Brachyura)	0.05	0.05	0.01	0	0.01	0.01	0.13
Stomatopods (Stomatopoda)	0	0.02	0.17	0	0	0	0.19
Bivalves (Bivalvia)	0	0	0.08	0	0	0	0.08
Gastropods (Gastropoda)	0	0	0	0	0	0	0
Squids (Decapodiformes)	0.01	0.04	0	0	0	0.01	0.06
Cuttlefish (Sepida)	0.01	0	0.11	0	0	0	0.12
Octopus (Octopoda)	0	0	0.11	0	0	0	0.11
Miscellaneous	0.36	0	0.85	0.3	0.13	0.07	1.71
Overall total	5.64	1.58	5.83	1.63	0.4	0.44	15.52

Source Department of Fisheries, States Government/UTs Administration; https://dof.gov.in/sites/default/files/2021-02/Final_Book.pdf

Table 2.13 Species-wise marine fish landings by coastal states and union territories: 2019–20 (in lakh tonnes) (West Coast)

Species	Goa	Gujarat	Karnataka	Kerala	Maharashtra	D & Diu, D & N Haveli	Lakshadweep	Total
Sharks (Selachimorpha)	0.01	0.09	0.01	0.01	0.08	0	0	0.2
Skates/Guitarfish (Rajidae)	0	0	0	0	0	0	0	0
Rays (Batoidea)	0	0	0	0.01	0	0	0	0.01
Eels (Anguilliformes)	0	0.03	0.01	0.01	0.02	0	0	0.07
Catfishes (Siluriformes)	0.01	0.27	0.03	0	0.09	0.01	0	0.41
Indian Oil sardine (*Sardinella longiceps*)	0.05	0	0.16	0.45	0.11	0	0	0.77
Other sardines (*Sardinella fimbriata* and others)	0.04	0	0.1	0.33	0	0	0	0.47
Hilsa shad (*Tenualosa ilisha*)	0	0.02	0.01	0	0.03	0	0	0.06
Other shads (Alosinae)	0	0	0.02	0	0.27	0	0	0.29
Coilia (*Coilia* spp.)	0	0.22	0.03	0	0	0.02	0	0.27
Setipinna (*Setipinna* spp.)	0	0	0.05	0	0	0	0	0.05
Stolephorus (*Stolephorus* spp.)	0	0	0	0.58	0	0	0	0.58
Thryssa (*Thryssa* spp.)	0	0	0.02	0.07	0.05	0	0	0.14
Wolf herring (*Chirocentrus* spp.)	0	0	0.02	0	0.04	0	0	0.06
Other clupeids	0	0.12	0.17	0.06	0.01	0	0	0.36
Bombay-duck (*Harpadon nehereus*)	0	0.89	0	0	0.32	0	0	1.21
Lizard fishes (*Synodontidae*)	0	0	0.12	0.17	0.04	0	0	0.33
Half beaks & full beaks (*Hemiramphidae & Belonidae*)	0	0	0.01	0.01	0	0	0	0.02
Flying fishes (*Exocoetidae*)	0	0	0	0	0	0	0	0
Rock cods (*Lotella rhacina*)	0	0	0	0.03	0	0	0	0.03

(continued)

Table 2.13 (continued)

Species	Goa	Gujarat	Karnataka	Kerala	Maharashtra	D & Diu, D & N Haveli	Lakshadweep	Total
Snappers (*Lutjanidae*)	0	0	0.07	0.01	0	0	0	0.08
Pig-face bream (*Lethrinidae*)	0	0	0.05	0	0.03	0	0	0.08
Threadfin bream (*Nemipteridae*)	0	0	0.32	0.28	0	0	0	0.6
Grouper (*Epinephelinae*)	0.01	0	0	0	0.01	0	0	0.02
Cobia (*Rachycentron canadum*)	0	0	0	0	0	0	0	0
Seabass (*Lates calcarifer*)	0	0	0	0	0	0	0	0
Other perches & perch like	0	0.11	0.28	0.08	0	0	0	0.47
Goatfishes (Mullidae)	0	0	0	0	0.11	0	0	0.11
Threadfins (Polynemidae)	0.01	0.04	0	0	0.03	0	0	0.08
Croakers (Sciaenidae)	0.02	1.33	0.08	0.05	0.05	0.02	0	1.55
Ribbon fishes (Trachipteridae)	0.01	0.5	0.18	0.05	0.27	0.18	0	1.19
Horse Mackerel (*Caranx & Trichiurus*)	0.01	0	0.07	0.02	0.11	0	0	0.21
Scads (*Decapterus* spp.)	0	0	0.2	0.16	0	0	0	0.36
Leather-jackets (*Oligoplites saurus*)	0.01	0.07	0.01	0	0	0.02	0	0.11
Pompano (*Trachynotus* spp.)	0	0	0	0	0	0	0	0
Other carangids	0.02	0.13	0.08	0.23	0	0	0	0.46
Silverbellies (Gerridae)	0.04	0.09	0.07	0.03	0	0	0	0.23
Big-jawed jumper (*Lactarius lactarius*)	0.01	0	0.06	0.01	0.04	0	0	0.12
Black pomfret (*Parastromatius niger*)	0	0.04	0.04	0.01	0.02	0.01	0	0.12
Silver pomfret (*Pampus argenteus*)	0.01	0.11	0.03	0.01	0.18	0.01	0	0.35
Indian mackerel (*Rastrelliger kanagurta*)	0.26	0	0.5	0.36	0.22	0	0	1.34
Other mackerels	0	0	0.07	0	0	0	0	0.07

(continued)

Table 2.13 (continued)

Species	Goa	Gujarat	Karnataka	Kerala	Maharashtra	D & Diu, D & N Haveli	Lakshadweep	Total
Scomberomorus spp.	0.02	0.12	0.04	0.04	0.11	0	0	0.33
Acanthocybium solandri	0	0	0	0	0	0	0	0
Kawa kawa (*Euthynnus affinis*)	0	0	0.11	0.05	0.14	0	0.01	0.31
Frigate tuna (*Auxis* spp.)	0.08	0	0	0	0	0	0.01	0.09
Skipjack tuna (*Katsuwonus pelamis*)	0	0	0	0.04	0	0	0.08	0.12
Longtail tuna (*Thunnus tonggol*)	0	0	0	0	0	0	0	0
Yellowfin tuna (*Thunnus albacares*)	0	0.07	0	0.07	0	0	0.05	0.19
Other tunnies	0	0	0	0	0	0	0.01	0.01
Bill fishes (Xiphiidae)	0	0	0.01	0.04	0	0	0	0.05
Barracudas (*Sphyraena* spp.)	0	0	0.05	0.03	0.05	0	0	0.13
Mullets (Mugilidae)	0	0.06	0	0	0	0	0	0.06
Unicorn cod (*Bregmaceros* mcclellandii)	0	0	0.13	0	0	0	0	0.13
Halibut (*Hippoglossus* spp.)	0	0	0.03	0	0	0	0	0.03
Flounders (Pleuronectoidei)	0	0	0.02	0	0	0	0	0.02
Soles (Soleidae)	0.01	0.11	0.05	0.14	0.09	0	0	0.4
Penaeid prawns (Penaeidae)	0.05	0.12	0.22	0.4	0.43	0	0	1.22
Non-penaeid prawns (Sergestidae)	0	0.41	0	0.03	0.49	0	0	0.93
Lobsters (Nephropidae)	0	0.02	0	0	0.01	0	0	0.03
Crabs (Brachyura)	0.01	0.06	0.03	0.05	0.04	0	0	0.19
Stomatopods (Stomatopoda)	0	0	0.19	0	0	0	0	0.19
Bivalves (Bivalvia)	0	0	0	0	0	0	0	0
Gastropods (Gastropoda)	0	0	0	0.02	0	0	0	0.02

(continued)

Table 2.13 (continued)

Species	Goa	Gujarat	Karnataka	Kerala	Maharashtra	D & Diu, D & N Haveli	Lakshadweep	Total
Squids (Decapodiformes)	0.02	0	0.17	0.18	0	0.03	0	0.4
Cuttlefish (Sepida)	0	0.36	0.07	0.12	0.32	0.01	0	0.88
Octopus (Octopoda)	0	0	0.02	0.05	0	0	0	0.07
Miscellaneous	0.3	1.62	0.02	0.46	0.62	0.01	0.04	3.07
Overall total	1.01	7.01	4.03	4.75	4.43	0.32	0.2	21.75

Source Department of Fisheries, States Government/UTs Administration; https://dof.gov.in/sites/default/files/2021-02/Final_Book.pdf

Table 2.14 State-wise fish consumption data (Per Capita/Year/Kg): 2019–20

S. No.	States/UT's	Yearly fish consumption (Per Capita/Kg) 2019–20
1	Andhra Pradesh	8.07
2	Arunachal Pradesh	3.52
3	Assam	11.72
4	Bihar	8.82
5	Chhattisgarh	4.66
6	Goa	NA
7	Gujarat	9.55
8	Haryana*	NA
9	Himachal Pradesh	2.16
10	Jharkhand	10.32
11	Karnataka	7.56
12	Kerala	19.41
13	Madhya Pradesh	2.76
14	Maharashtra	3.02
15	Manipur	14.1
16	Meghalaya	10.98
17	Mizoram	5.54
18	Nagaland	6.68
19	Odisha	13.79
20	Punjab	0.4
21	Rajasthan	0.01
22	Sikkim	1.16
23	Tamil Nadu	9.6
24	Telangana	8.87
25	Tripura	29.29
26	Uttarakhand	0.49
27	Uttar Pradesh	10.89
28	West Bengal	NA
29	A and N Islands	59.47
30	Chandigarh	NA
31	Daman and Diu, D & Nagar Haveli	NA
32	Delhi	NA
33	Jammu & Kashmir	3
34	Ladakh	NA
35	Lakshadweep	NA
36	Puducherry	30

Source Department of Fisheries, States Government/UTs Administration; https://dof.gov.in/sites/default/files/2021-02/Final_Book.pdf

Table 2.15 Annual average fish landings by species in the Hooghly estuary (in metric tons). Figures in parentheses indicate percentage of the catch. (*) less than 0.1% of catch

Species	Pre-Farakka period (1966–1967 to 1974–1975)	Post-Farakka period (1984–1985 to 1989–1990)	Our observation[#] (2021–2022)
Tenualosa ilisha (Family Clupeidae)	1,077.10 (14.4)	1,017.70 (3.6)	1,987.55 (4.47)
Setipinna spp. Family Engraulidae)	651.1 (8.7)	3,243.50 (11.6)	4,658.24 (10.47)
Harpodon nehereus (Family Harpodontidae)	1,929.10 (25.8)	4,932.30 (17.7)	7855.56 (17.65)
Trichiurus spp. (Family Trichiuridae)	452 (6)	2,996.50 (10.7)	5102.91 (11.46)
Pama pama (Family Sciaenidae)	172.7 (2.3)	3,700.40 (13.3)	4500.75 (11.11)
Silago panijus (Family Sillaginidae)	17.4 (0.2)	15.3 (*)	29.24 (0.07)
Tachysurus jella (Family Ariidae)	176 (2.3)	597.7 (2.1)	728.75 (1.64)
Polynemus paradiseus (Family Polynemidae)	18.6 (0.2)	91.6 (0.3)	115.56 (0.26)
Coilia spp. (Family Engraulidae)	70.9 (0.9)	800.3 (2.9)	1024.66 (2.30)
Tenualosa toli (Family Clupeidae, SF Alosinae)	16.9 (0.2)	49.2 (0.2)	56.88 (0.13)
llisha elongata (Family Pristigasteridae)	164.2 (2.2)	447.3 (1.6)	598.05 (1.34)
Eleutheronema tetradactylum (Family Polynemidae)	24.8 (0.3)	14.6 (*)	65.72 (0.15)
Sciaena biauritus (Family Sciaenidae)	188.8 (2.5)	303.3 (1.1)	415.98 (0.93)
Pangasius pangasius (Family Pangasiidae)	76.1 (1)	3.7 (*)	123.20 (0.28)
Liza parsia (Family Mugilidae)	42.9 (0.6)	18 (0.1)	54.75 (0.12)
Lates calcarifer (Family Centropomidae)	24.1 (0.3)	6.2 (*)	45.98 (0.10)
Pampus argenteus (Family Stromateidae)	–	452.5 (1.6)	789.15 (1.77)
Prawns	751.1 (10.1)	1,856.60 (6.6)	3571.63 (8.02)
Freshwater fishes	–	–	–

(continued)

Table 2.15 (continued)

Species	Pre-Farakka period (1966–1967 to 1974–1975)	Post-Farakka period (1984–1985 to 1989–1990)	Our observation# (2021–2022)
Others	1609.3 (21.6)	7,351.30 (26.3)	12,786.95 (28.73)
Total	7463.1	27,898	44,511.51

Source Sinha et al. (1996); '#' our field-based data (mean) collected from nine fish landing stations

Fig. 2.15 Shrimp farms in Sundarbans at the cost of mangroves

study on the socio-economic profile of cow, goat, and sheep farmers at Bali II Island (22°05′44.1″N; 88°46′38.1″E) during May, 2023 in central Indian Sundarbans. The survey report is highlighted in Table 2.16.

The products provided by Sundarban livestock like meat, milk, eggs, skin etc. have both internal island market as well external markets like in the city of Kolkata, Howrah etc., which are adjacent to the northern border of Indian Sundarbans. A respondent analysis carried out on the provisioning services of livestock sector by involving the major stakeholders in the western and central Indian Sundarbans during 2022 exhibit the inclination of the respondents towards meat followed by milk, egg, manure, and skin (Tables 2.17 and 2.18).

Our survey consisted of few steps: The first round of questions was prepared and an online invitation (through google meet) for participation was sent to the selected respondents ($n > 50$) (Annexure 2.1). The first-round questions were open-ended and the respondents were asked to suggest major Provisioning Services (PS) provided by livestock of Indian Sundarbans based on their age-old experience. A feedback

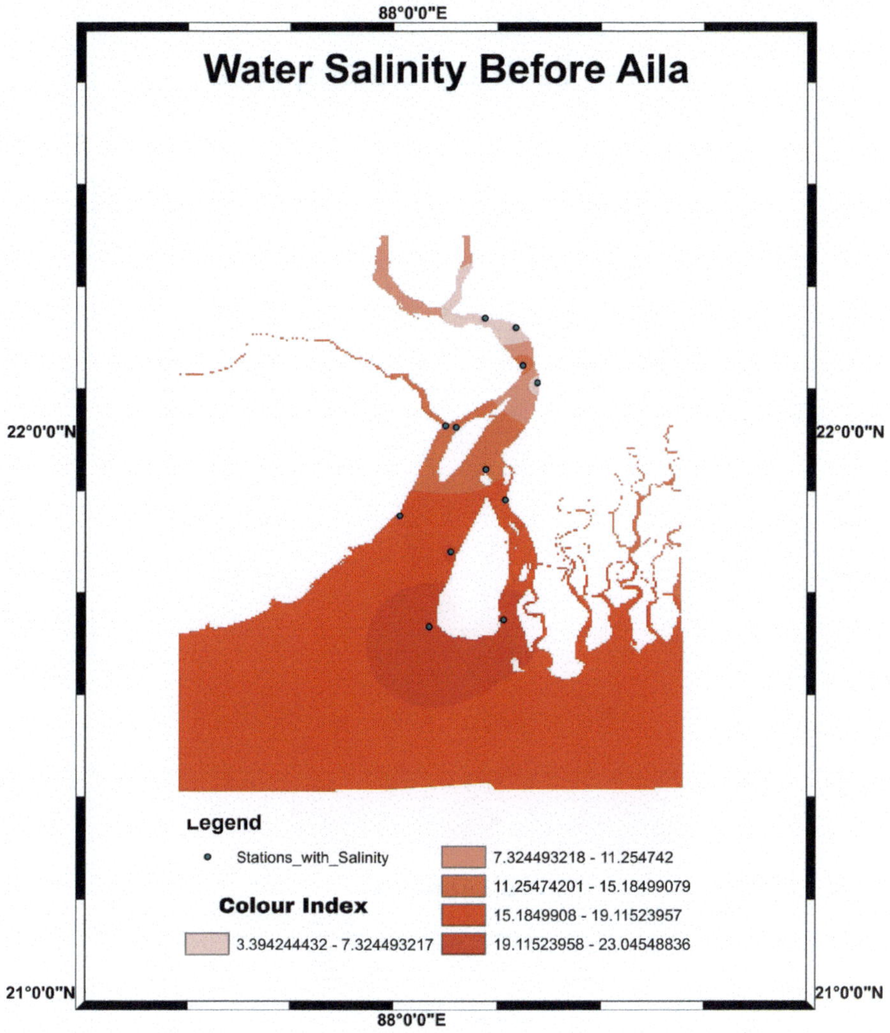

Fig. 2.16 Mean aquatic salinity (in psu) in 12 selected stations during pre-Aila

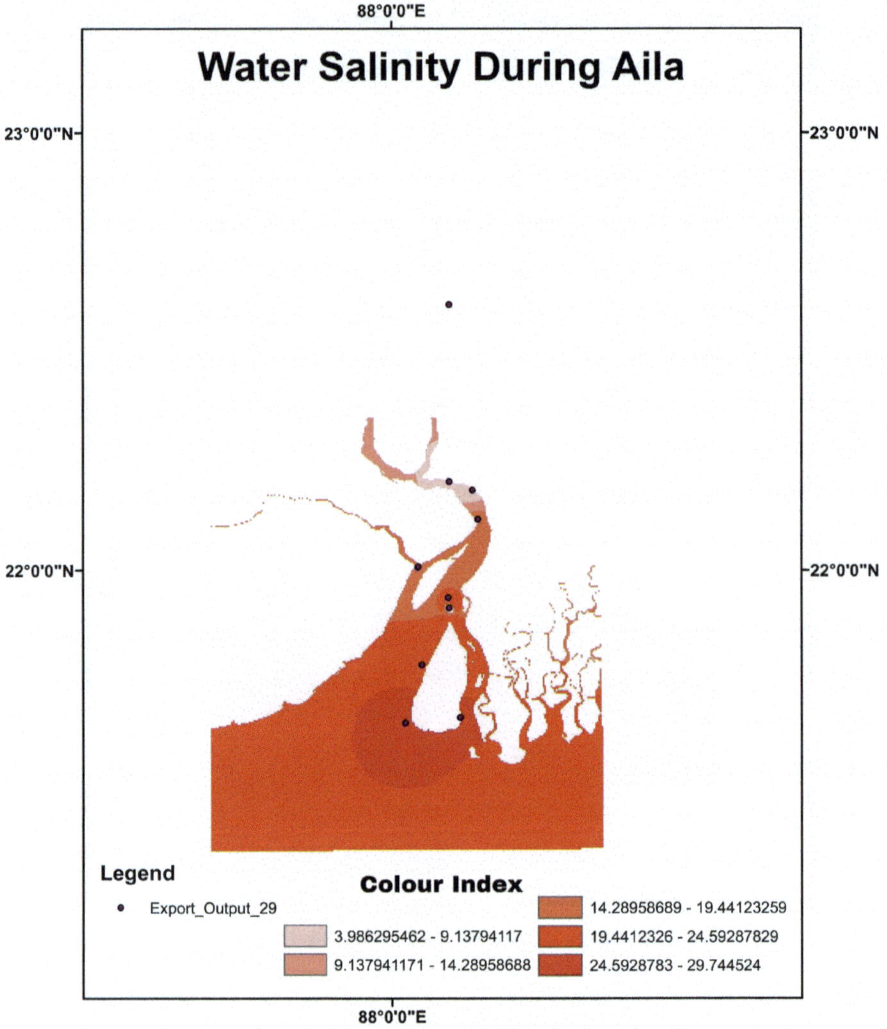

Fig. 2.17 Mean aquatic salinity (in psu) in 12 selected stations during Aila

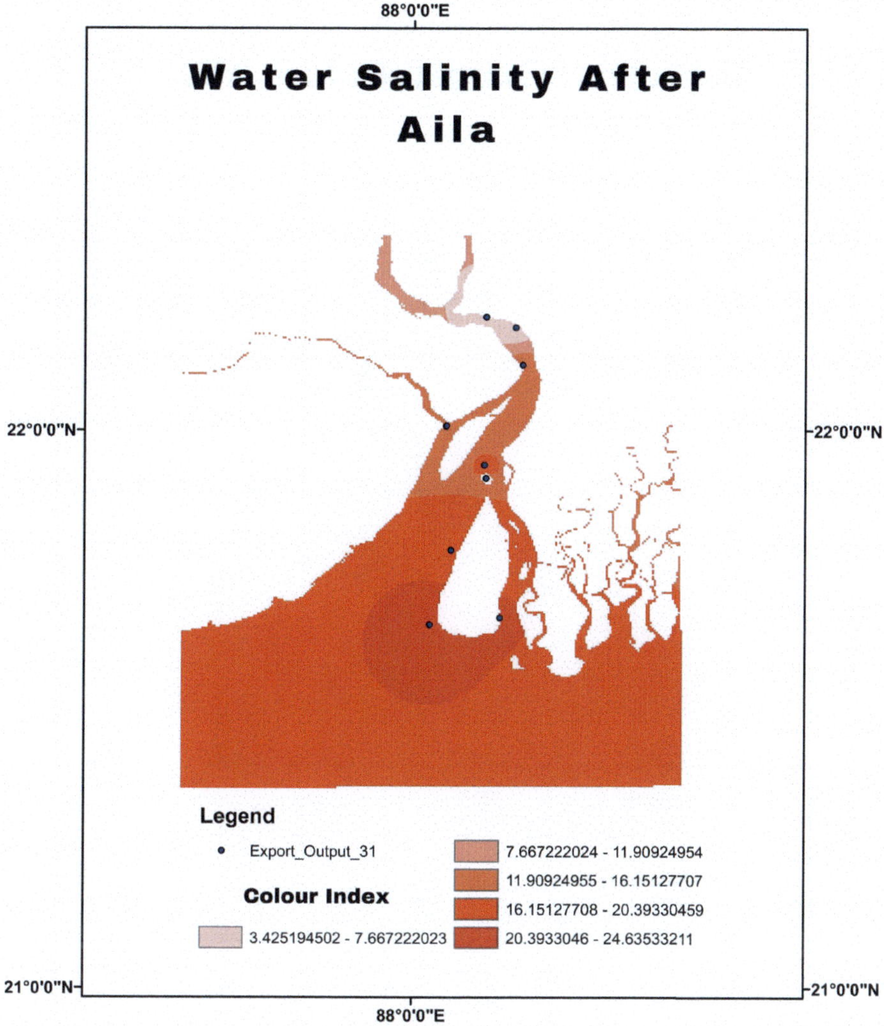

Fig. 2.18 Mean aquatic salinity (in psu) in 12 selected stations during post-Aila

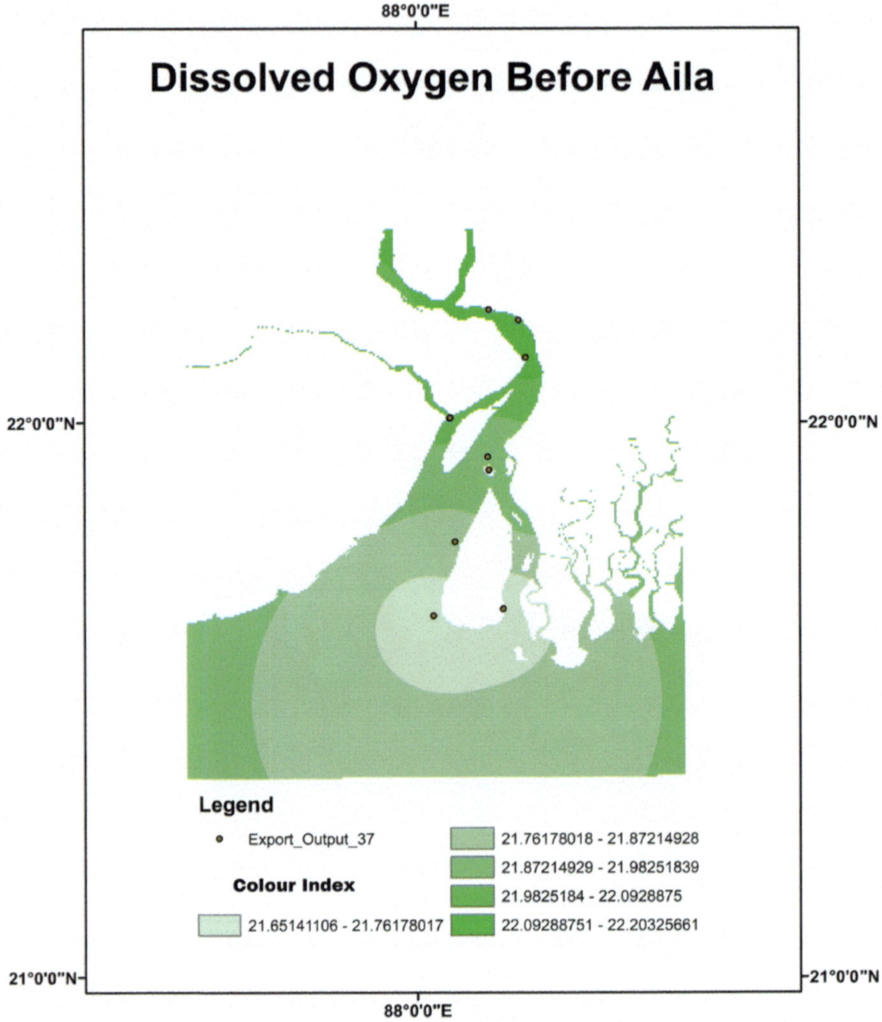

Fig. 2.19 Mean disolved oxygen (in ppm) in 12 selected stations during pre-Aila

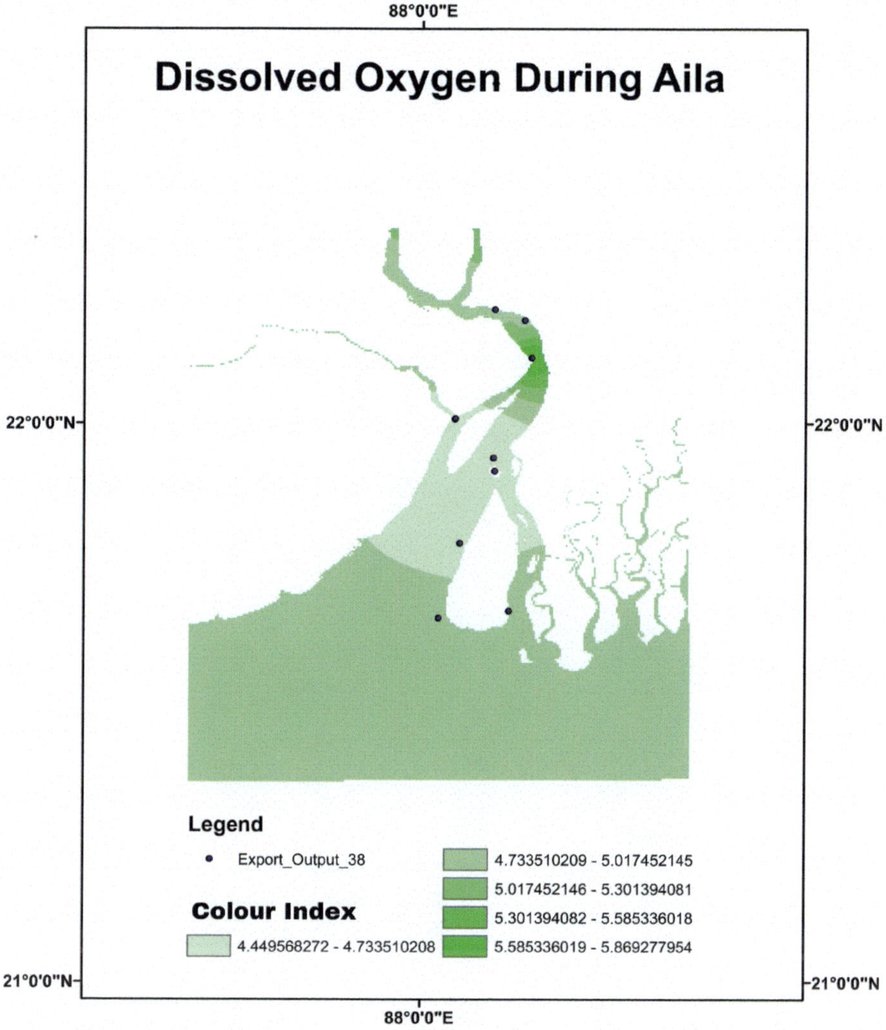

Fig. 2.20 Mean disolved oxygen (in ppm) in 12 selected stations during Aila

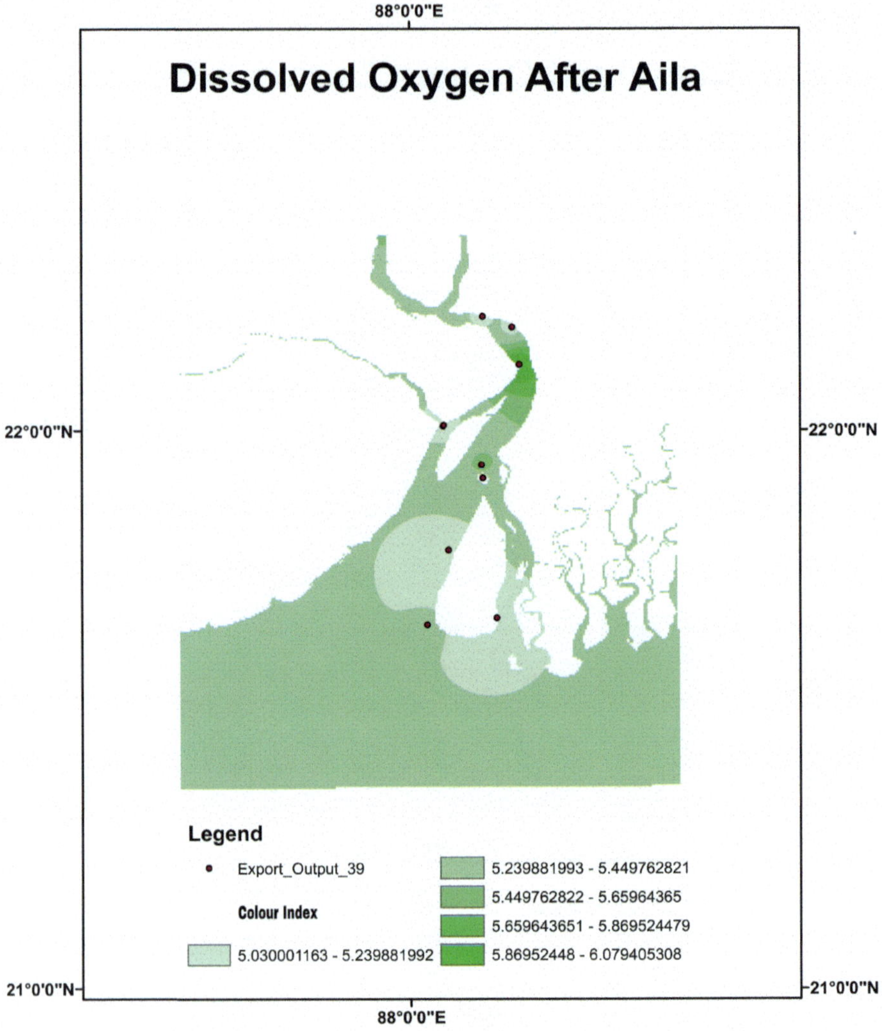

Fig. 2.21 Mean disolved oxygen (in ppm) in 12 selected stations during post-Aila

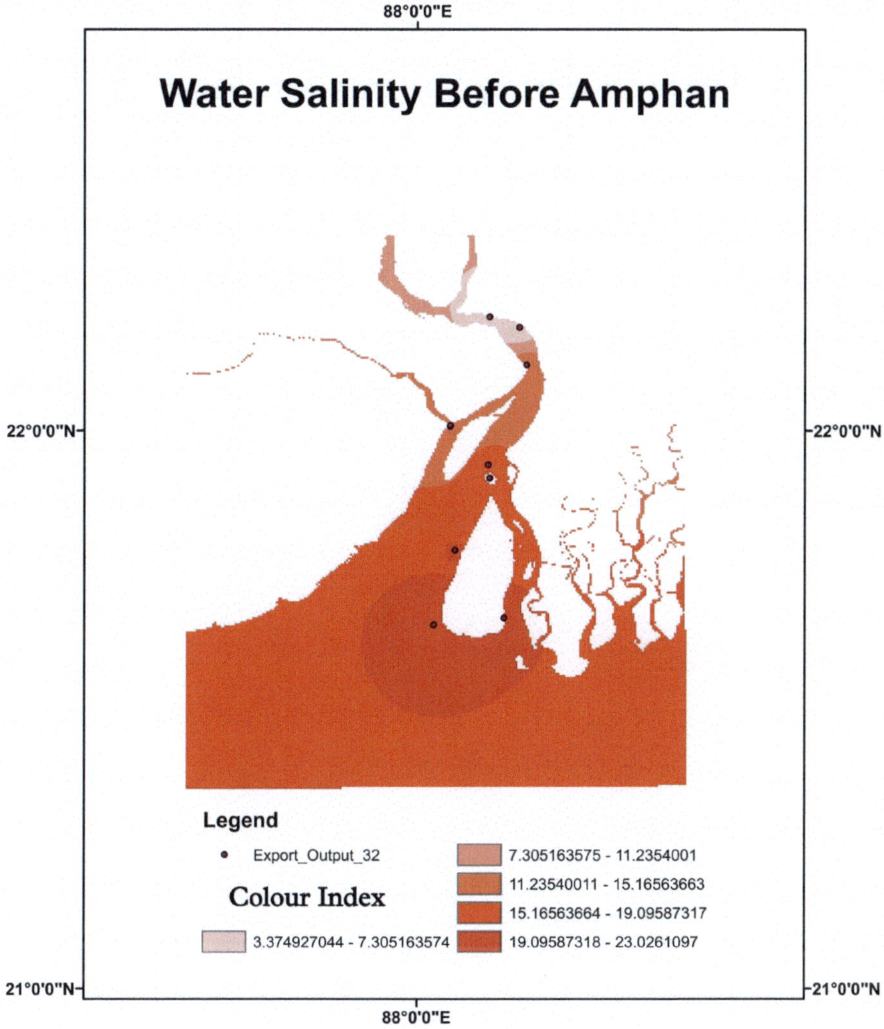

Fig. 2.22 Mean aquatic salinity (in psu) in 12 selected stations during pre-Amphan

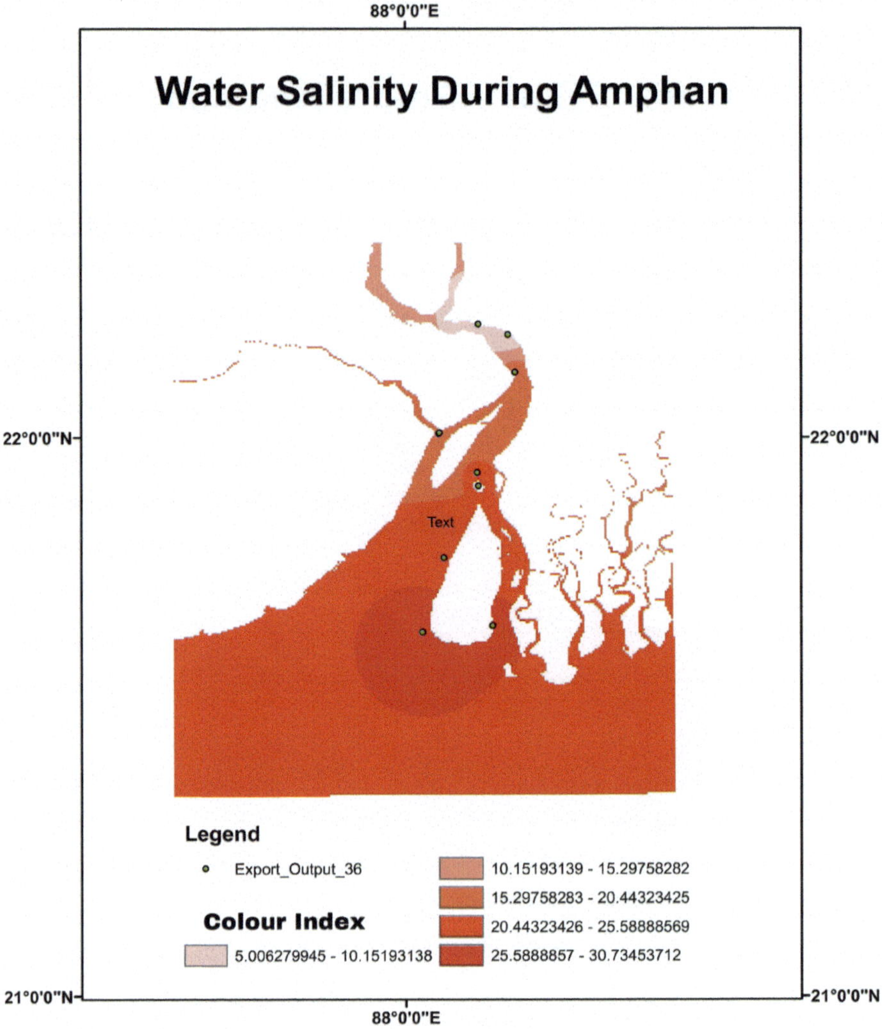

Fig. 2.23 Mean aquatic salinity (in psu) in 12 selected stations during Amphan

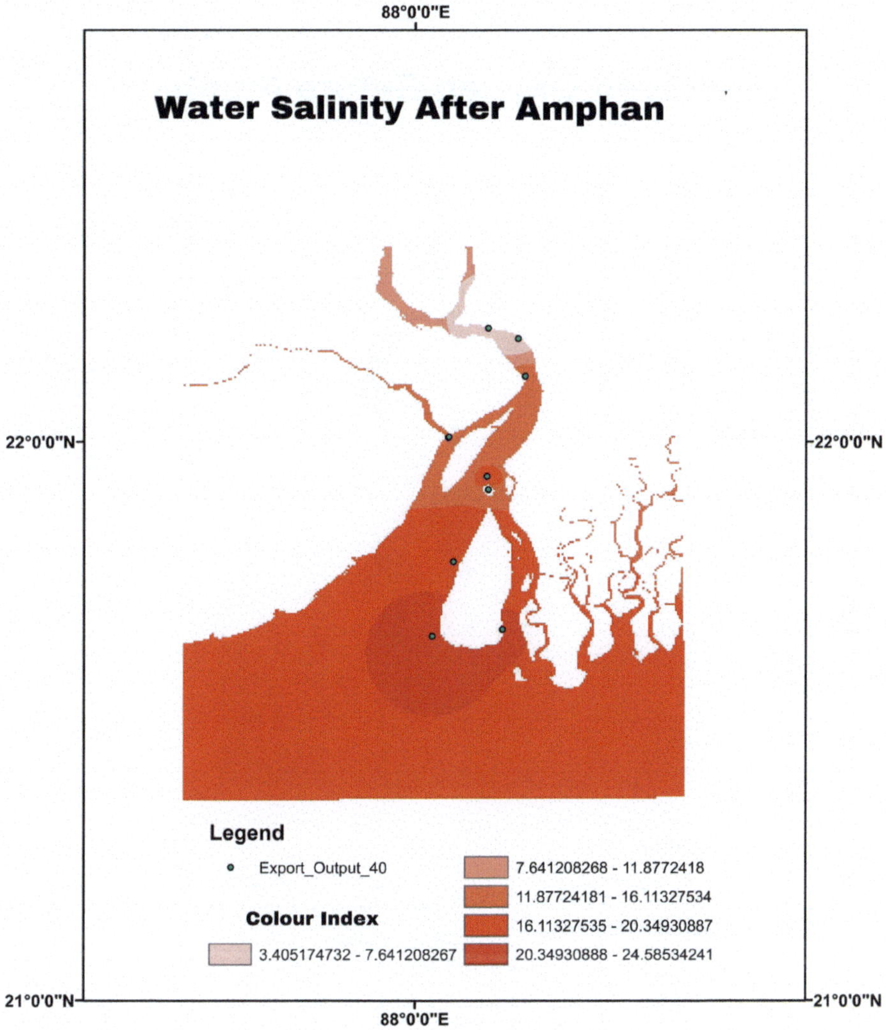

Fig. 2.24 Mean aquatic salinity (in psu) in 12 selected stations after Amphan

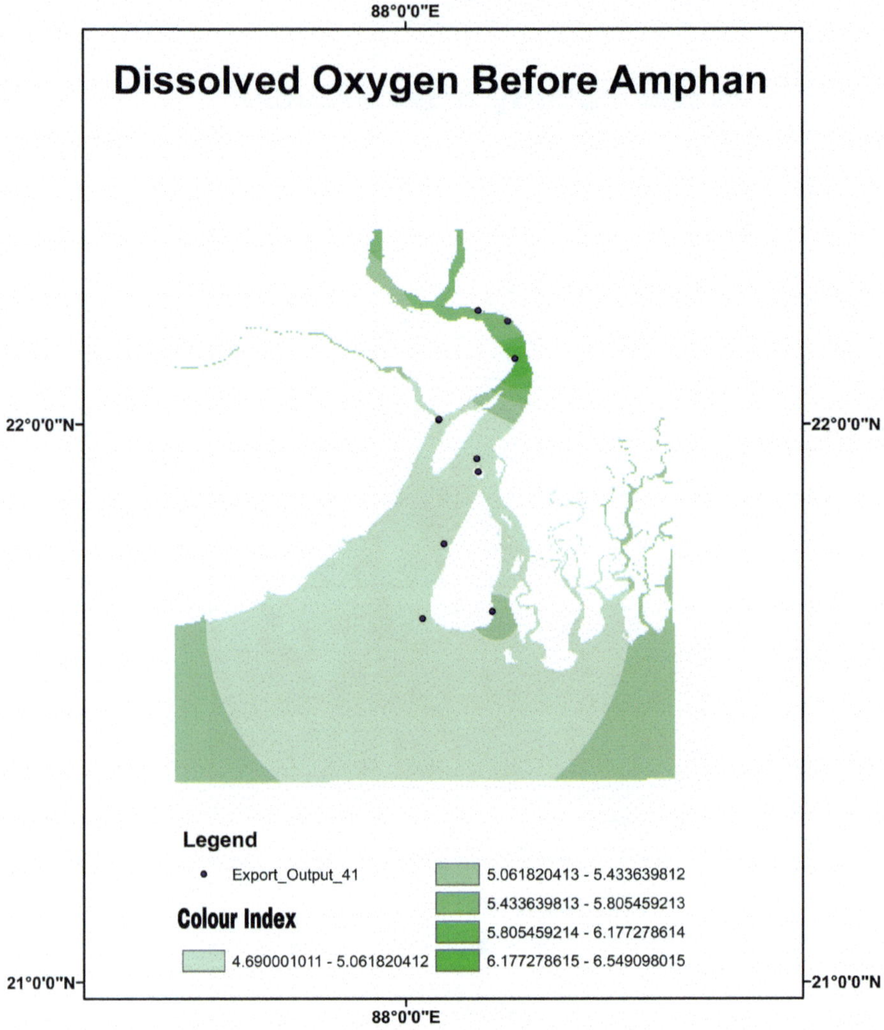

Fig. 2.25 Mean disolved oxygen (in ppm) in 12 selected stations during pre-Amphan

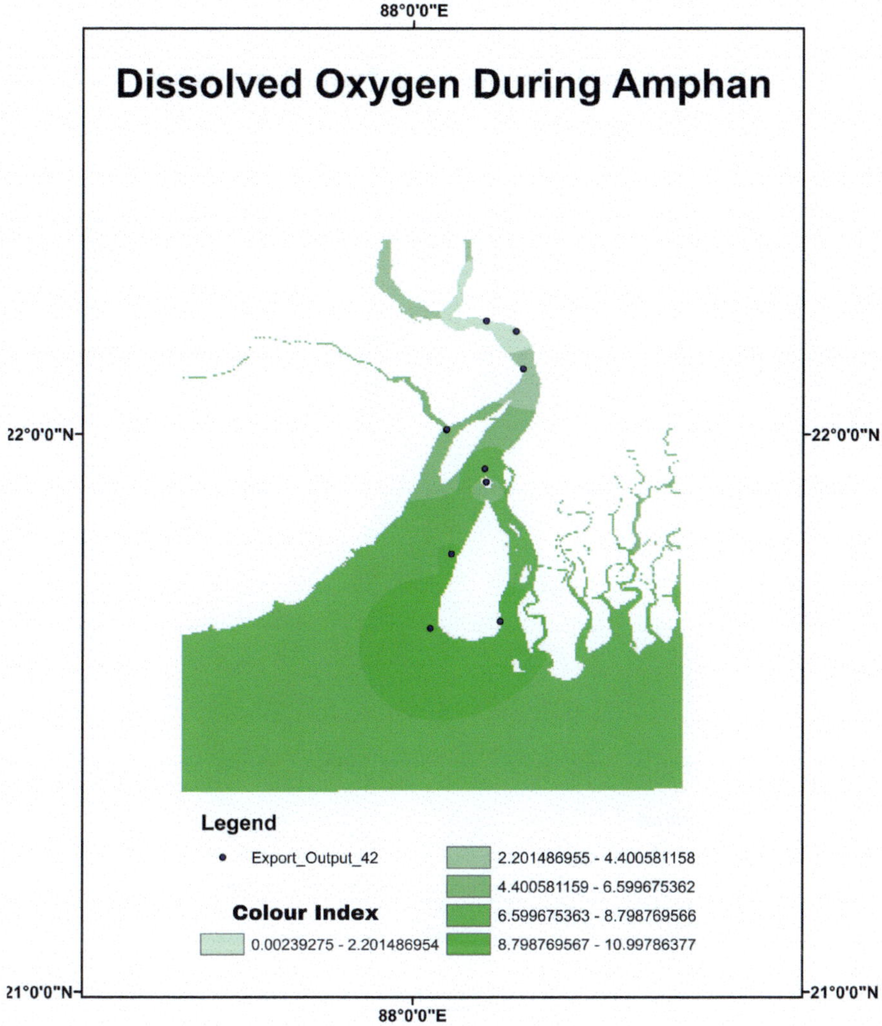

Fig. 2.26 Mean disolved oxygen (in ppm) in 12 selected stations during Amphan

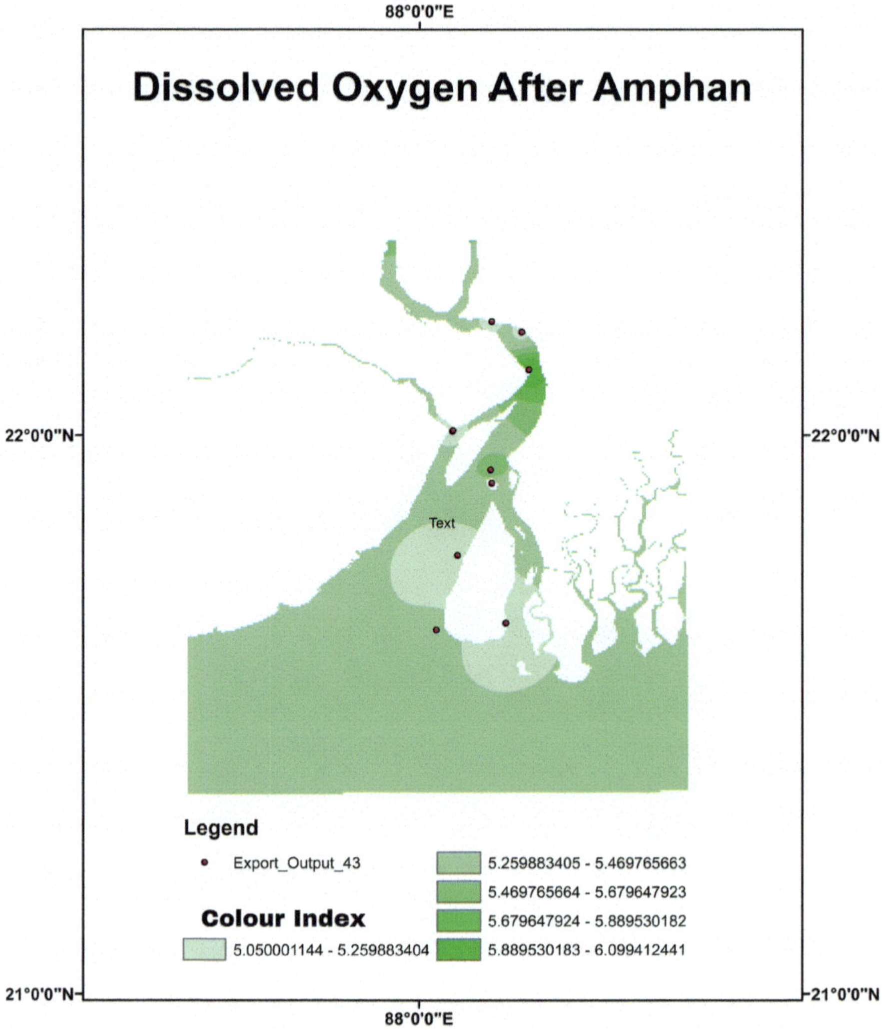

Fig. 2.27 Mean disolved oxygen (in ppm) in 12 selected stations during post-Amphan

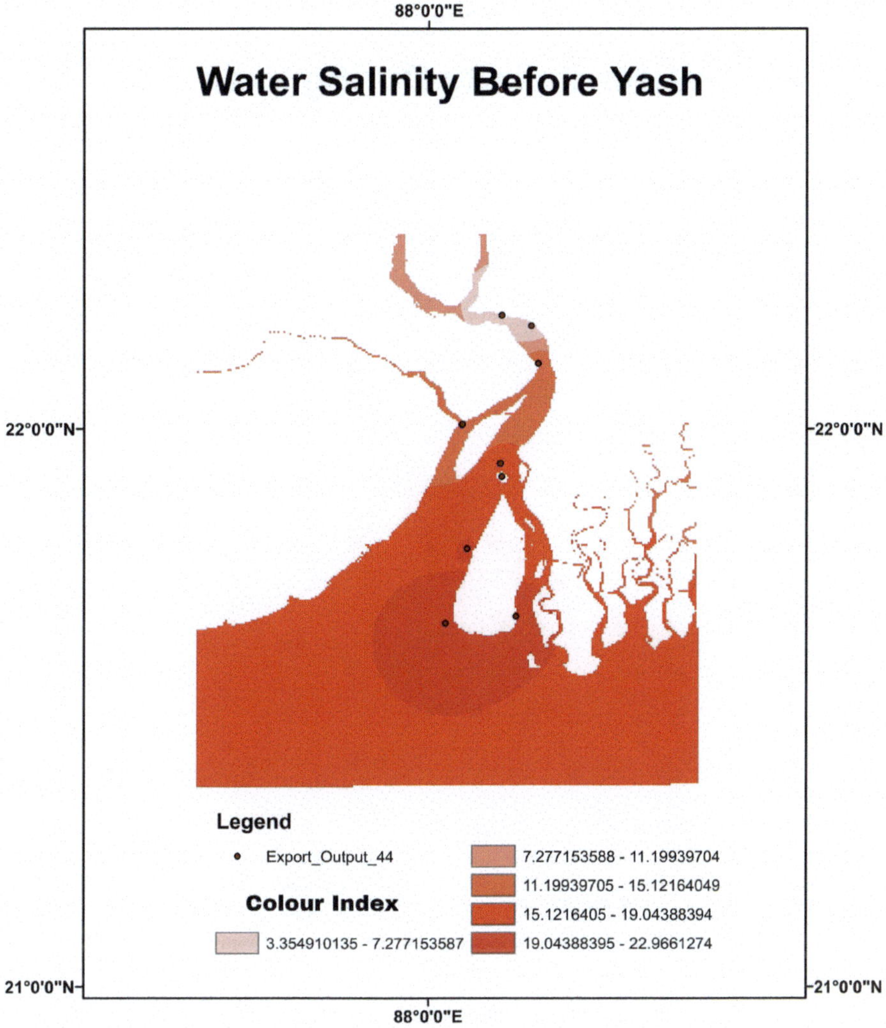

Fig. 2.28 Mean aquatic salinity (in psu) in 12 selected stations during pre-Yash

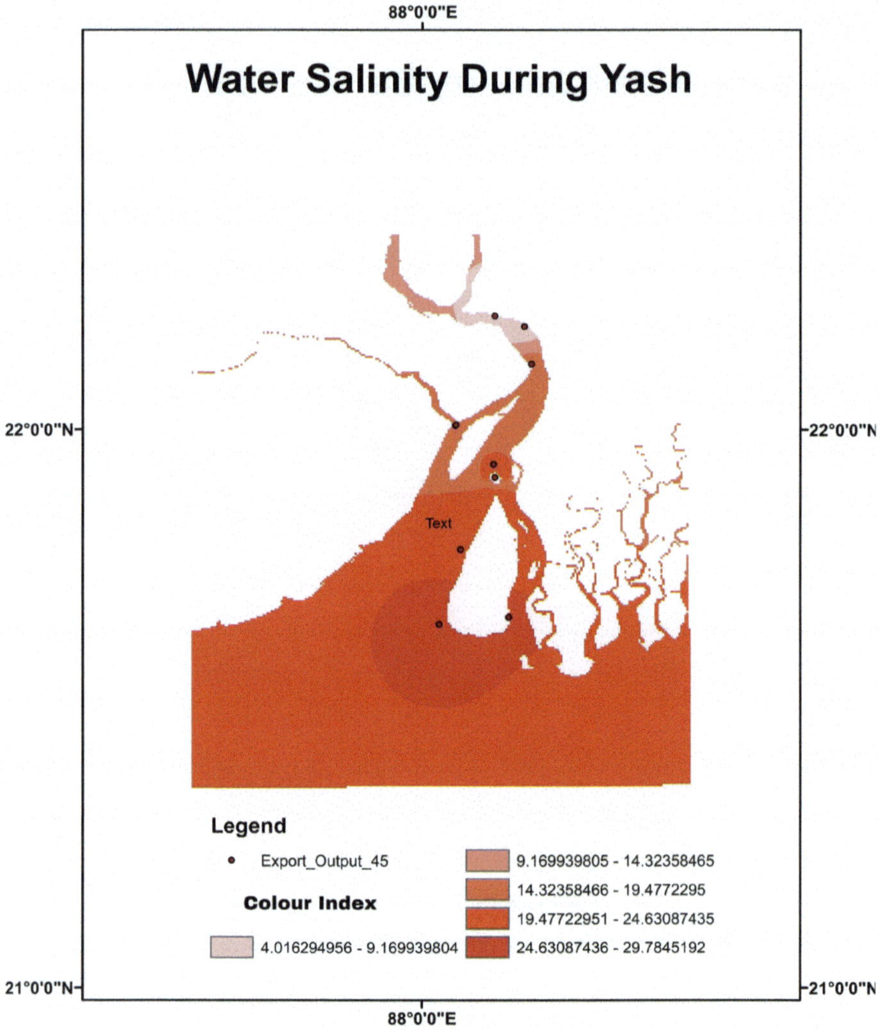

Fig. 2.29 Mean aquatic salinity (in psu) in 12 selected stations during Yash

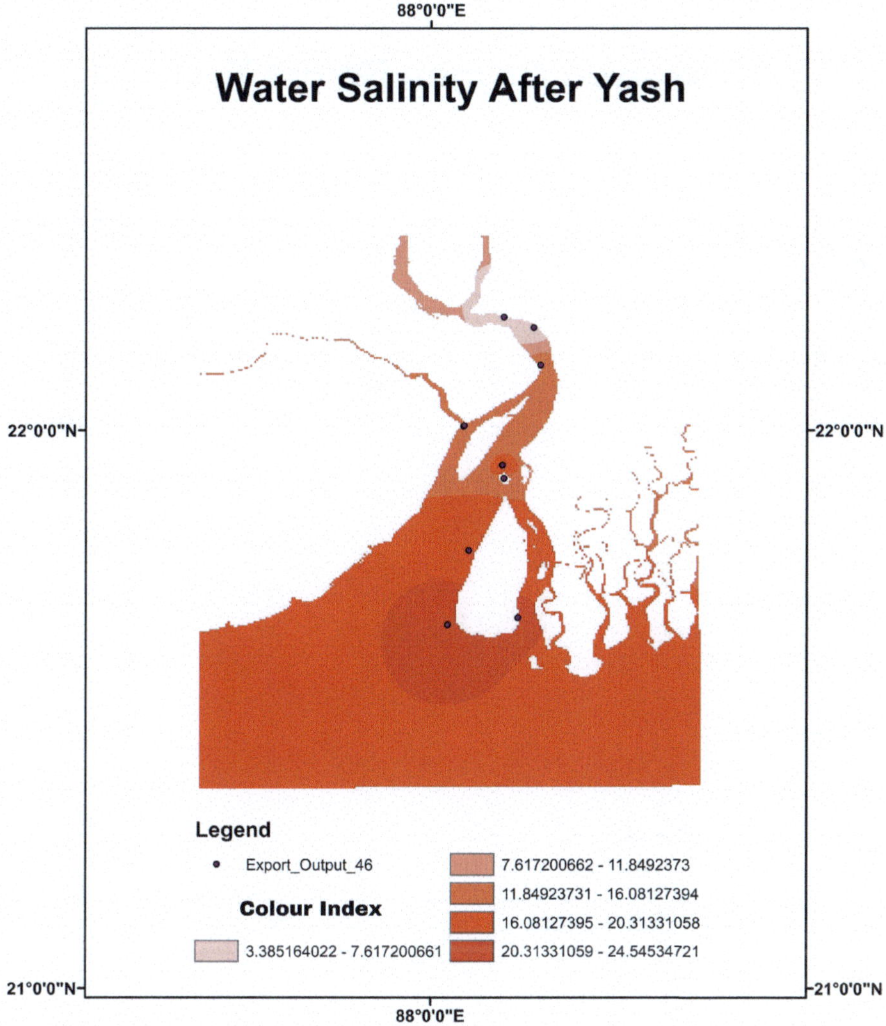

Fig. 2.30 Mean aquatic salinity (in psu) in 12 selected stations during post-Yash

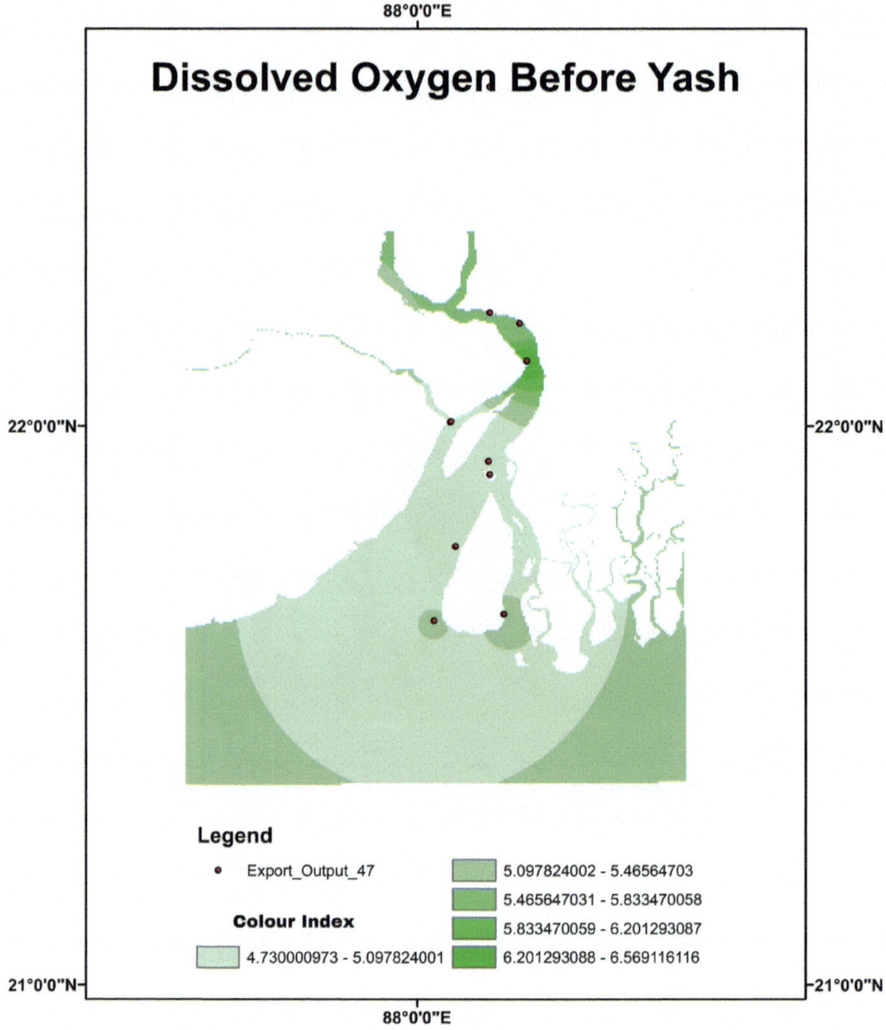

Fig. 2.31 Mean disolved oxygen (in ppm) in 12 selected stations during pre-Yash

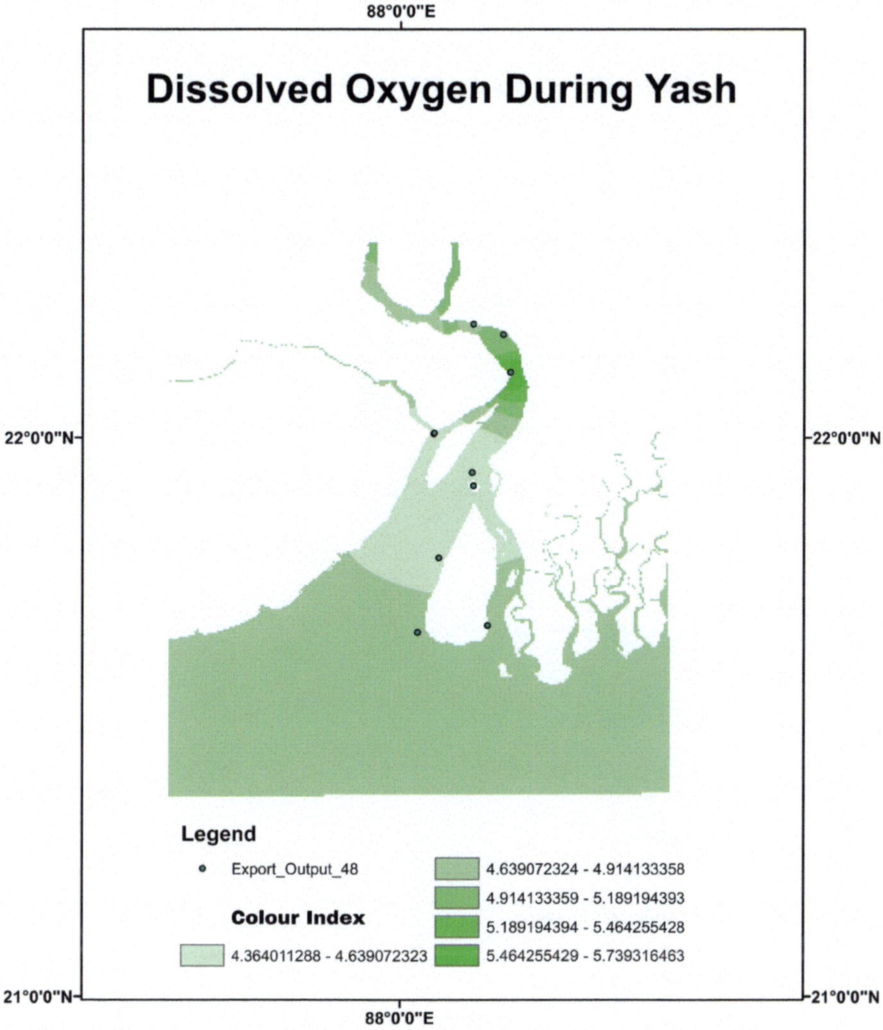

Fig. 2.32 Mean disolved oxygen (in ppm) in 12 selected stations during Yash

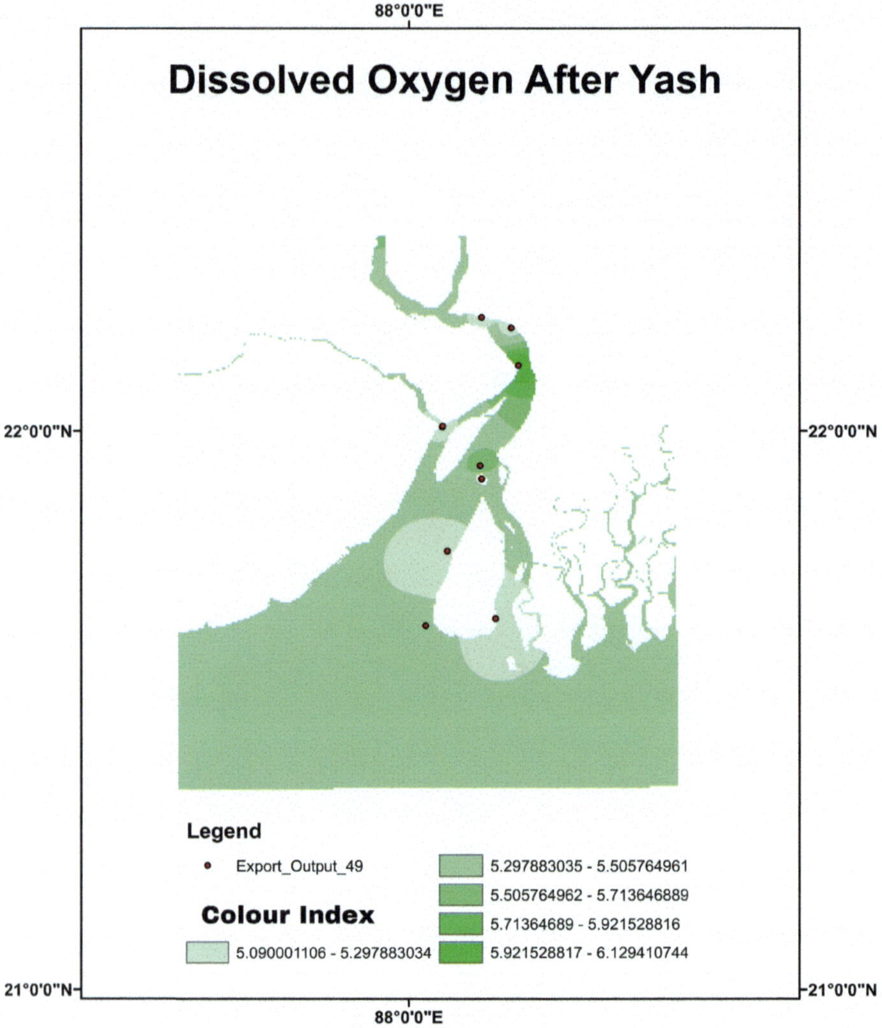

Fig. 2.33 Mean disolved oxygen (in ppm) in 12 selected stations during post-Yash

Fig. 2.34 Goat rearing in Indian Sundarbans is a profitable venture in the domain of traditional livelihood

Fig. 2.35 Rearing of pigs provide money to the farmers as they are sold in the 'haat' (local market in the rural areas)

Fig. 2.36 White pigs are reared in Indian Sundarbans because of their high growth rate, frequency of breeding and demand for pork

Table 2.16 Socio-economic profile of cow, goat, and sheep farmers (n = 1254)

Characters	Category	No.	Percentage
Gender	Male	2055	34.11
	Female	3970	65.89
Age	People between the age group 20–30 yrs	1719	28.53
	People between the age group 30–60 yrs	2925	48.55
	People between the age group 60–70 yrs	1381	22.92
Occupation	Agriculturist	2604	43.22
	Fisherman	135	2.24
	Businessman	2282	37.88
	Daily labour	63	1.04
	Service	941	15.62
Education of the respondent	Class VIII pass	3091	51.3
	Secondary	1295	21.5

(continued)

Table 2.16 (continued)

Characters	Category	No.	Percentage
	Higher secondary	1422	23.6
	Graduate and higher	217	3.6
Religion	Hindu	3931	65.25
	Muslim	2055	34.1
	Christian	39	0.65
Capacity building and requisite training	No training	4534	75.25
	Partial training	905	15.02
	Full training	586	9.73
Family income per annum	INR up to 30,000	693	11.5
	INR 30,000–50,000	4117	68.33
	INR 50,000 and above	1215	20.17

Table 2.17 Major provisioning services offered by livestock in western Indian Sundarbans

Provisioning service	Policy formulator (respondent type 1)		
	PSR	% of vote	PSS 1
Meat	5	33.7	168.5
Milk	3	27.8	83.4
Egg	2	19.8	39.6
Skin	1	6.9	6.9
Manure	3	11.8	35.4
Provisioning service	Cattle farmer (respondent type 2)		
	PSR	% of vote	PSS 2
Meat	5	34.1	170.5
Milk	3	25.4	76.2
Egg	3	17.4	52.2
Skin	1	5.9	5.9
Manure	2	17.2	34.4
Provisioning service	Fisherman (respondent type 3)		
	PSR	% of vote	PSS 3
Meat	5	32.7	163.5
Milk	3	24.9	74.7
Egg	3	20.8	62.4
Skin	2	5.7	11.4
Manure	2	15.9	31.8

(continued)

Table 2.17 (continued)

Provisioning service	Agriculturist (respondent type 4)		
	PSR	% of vote	PSS 4
Meat	5	36.7	183.5
Milk	4	23	92
Egg	4	15.6	62.4
Skin	1	6.9	6.9
Manure	3	17.8	53.4
Provisioning service	Daily labour (respondent type 5)		
	PSR	% of vote	PSS 5
Meat	5	34.1	170.5
Milk	4	24.2	96.8
Egg	3	19.7	59.1
Skin	1	6.7	6.7
Manure	3	15.3	45.9

Table 2.18 Major provisioning services offered by livestock in central Indian Sundarbans

Provisioning service	Policy formulator (respondent type 1)		
	PSR	% of vote	PSS 1
Meat	5	35	175
Milk	3	26.4	79.2
Egg	2	19.8	39.6
Skin	1	7.3	7.3
Manure	3	11.5	34.5
Provisioning service	Cattle farmer (respondent type 2)		
	PSR	% of vote	PSS 2
Meat	5	36.7	183.5
Milk	3	25.6	76.8
Egg	3	17.4	52.2
Skin	1	4.7	4.7
Manure	2	15.6	31.2
Provisioning service	Fisherman (respondent type 3)		
	PSR	% of vote	PSS 3
Meat	5	33.7	168.5
Milk	3	23.7	71.1
Egg	3	19.7	59.1
Skin	1	6.3	6.3

(continued)

Table 2.18 (continued)

Provisioning service	Fisherman (respondent type 3)		
	PSR	% of vote	PSS 3
Manure	2	16.6	33.2
Provisioning service	Agriculturist (respondent type 4)		
	PSR	% of vote	PSS 4
Meat	5	37.8	189
Milk	4	21.4	85.6
Egg	3	15.6	46.8
Skin	1	7.3	7.3
Manure	2	17.9	35.8
Provisioning service	Daily labour (respondent type 5)		
	PSR	% of vote	PSS 5
Meat	5	36.7	183.5
Milk	3	23.7	71.1
Egg	3	19.7	59.1
Skin	1	5.2	5.2
Manure	3	14.7	44.1

report was prepared based on the survey. In the second round, scores were provided to these categories of PS, where the respondents were asked to score options on a scale of 1–5, where "1" indicates lowest value and "5" indicates highest value. Later, to rank the categories, the scores for each PS category was given corresponding weights ranging between 1 to 5 (PSR) and multiplied by the percentage of votes for that option to generate a total weighted score for that PS, which is referred to as Provisioning Service Score (PSS).

In the final stage, Composite Provisioning Service Scale (CPSS) was constructed based on Provisioning Service Scale (PSS) computed as per the expression:

$$\text{CPSS} = \text{PSS1} + \text{PSS2} + \text{PSS3} + \text{PSS4} + \text{PSS5}$$

where, PSS = Provisioning Service Rank (PSR) × % of Vote.

It is to be noted in this context that the sample size of different respondents is variable e.g., the sample size for policy formulator is not like daily labor or fisherman. For policy formulator, we could involve 57 members, but for other respondents n = 155.

Considering the Composite Provisioning Service Scale (CPSS), the trend of livelihood products in western Indian Sundarbans is meat (856.5) > milk (423.1) > egg (275.7) > manure (200.9) > Skin (37.8).

In context to Composite Provisioning Service Scale (CPSS), the trend of livelihood products in central Indian Sundarbans is meat (899.5) > milk (383.8) > egg (256.8) > manure (178.8) > skin (30.8).

India is presently occupying 3rd position in terms of egg production, producing some 70 billion eggs per annum. Most of these eggs are procured from the rural areas of the countries where the traditional Back Yard Poultry Farming (BYPF) contributes significantly in the country's egg and meat production. Government of India has also given priority to scale-up this livelihood domain for a steady supply of animal protein and employment generation in rural areas (GoI 2005; 2008). In Indian Sundarbans a major percentage of the population depends on poultry farming for sustaining their livelihoods. Based on the climatic condition of the area, majority of the poultry farmers rear 'desi' poultry birds (74.1%), Rhode Island Red (12.8%) and Kroiler (13.1%). Majority of the poultry farmers in Indian Sundarbans belong to Low Income Group earning about INR 4,000–6,000 per month.

There are also farmers who are involved in large-scale farming and earn about INR 1,50,00–2,00,00 per month. We made a survey during 2021–22 in Chotomollakhali Island (22°10′40.00″N and 88°54′26.71″E) located in central Indian Sundarbans to acquire the realistic picture of socio-economic status of poultry farmers in this region (Table 2.19).

Performance of BYPF can be improved through proper management of the livestock like vaccination, de-worming, use of clean water, application of sanitizer in the

Table 2.19 Socio-economic profile of poultry farmers in Chotomollakhali Island (22°10′40.00″N and 88°54′26.71″E) under central Indian Sundarbans

Variables	Category	Percentage
Age	15–20 years	2.5
	21–45 years	73.8
	46–70 years	23.7
Sex	Male	23.8
	Female	76.2
Education	Class I-VI	44.5
	Class VII-VIII	30.9
	Class VIII—Secondary level	20.2
	Higher Secondary and above	4.8
Religion	Hindu	68.4
	Islam	23.2
	Christian	8.4
Farming experience	1–3 years	6.2
	4–10 years	12.8
	More than 10 years	81.0
Gross income from BYPF per month	INR 500–1,000/-	34.2
	INR 1,000–2,500/-	41.8
	INR 2,500–4,000/-	14.6
	INR 4,000–10,000/-	9.4

Table 2.20 Temporal variation of avian disease in Indian Sundarbans

Year	% of avian influenza
2011	33.5
2012	36.1
2013	30.2
2014	28.4
2015	31.3
2016	26.3
2017	23.7
2018	21.2
2019	19.5
2020	21.3
2021	14.3
2022	11.9

COVID-19 like scenario etc. We have found from our survey that about 67.5% of the poultry farmers use better management practices and focus on sufficient health care measures. This has reduced the percentage of avian influenza in the region (Table 2.20).

Duck rearing is also a productive livestock business in the globe because of its egg and meat. Sometimes the feathers of ducks are used to make cleaners and decorative items. Duck eggs are relatively larger, weighing about 4.6% of duck's body weight, compared to chicken, whose egg weight is only about 3.4% of the hen's body weight (Narhari 2009).

Most of the duck owners were cultivators (45.45%) followed by labourers (26.20%), service class (17.12%), business class (3.55%) and others (7.68%). In Indian Sundarbans, duck farmers were mostly small land holder (47.43%) followed by landless farmers (27.23%), marginal farmers (18.68%) and medium-large land holding farmers (6.66%). Ducks were mostly reared by females in all sectors of Indian Sundarbans.

Considering the major traditional livelihoods of the inhabitants of Indian Sundarbans, we carried out a survey during 2022 in the western and central sectors to monitor the preference of the respondents towards the existing livelihoods (Tables 2.21 and 2.22).

Our survey consisted of few steps: The first round of questions was prepared and an online invitation (through google meet) for participation was sent to the selected respondents (n > 100). The first-round questions were open-ended and the respondents were asked to suggest major Livelihood Preference (LP) amongst agriculture, pisciculture, aquaculture (preferable shrimp culture), cow rearing, goat rearing, pig rearing and poultry (Total = 7 categories). A feedback report was prepared based on the survey. In the second round, scores were provided to these categories of LP, where the respondents were asked to score options on a scale of 1–7, where "1" indicates lowest preference and "7" indicates highest preference. Later, to rank the categories,

Table 2.21 Major Livelihood Preference (LP) of respondents in western Indian Sundarbans

Livelihood	Policy formulator (respondent type 1)		
	LPR	% of vote	LPS1
Agriculture	5	20.4	102
Pisciculture and deep-sea fishing	7	35.1	245.7
Aquaculture (preferably shrimp farming)	6	27.8	166.8
Cow rearing	4	4.8	19.2
Goat rearing	4	5.8	23.2
Pig rearing	1	1.5	1.5
Poultry	3	4.6	13.8
	Agriculturist (respondent type 2)		
	LPR	% of vote	LPS2
Agriculture	5	21.4	107
Pisciculture and deep-sea fishing	6	33.8	202.8
Aquaculture (preferably shrimp farming)	5	28.3	141.5
Cow rearing	3	4.6	13.8
Goat rearing	4	5.1	20.4
Pig rearing	1	1	1
Poultry	2	3.6	7.2
	Fisherman (respondent type 3)		
	LPR	% of vote	LPS3
Agriculture	4	19	76
Pisciculture and deep-sea fishing	7	38.7	270.9
Aquaculture (preferably shrimp farming)	6	32.5	195
Cow rearing	3	2.4	7.2
Goat rearing	5	3	15
Pig rearing	1	1	1
Poultry	3	1.9	5.7
	Livestock owner (respondent type 4)		
	LPR	% of vote	LPS4
Agriculture	5	9.1	45.5
Pisciculture and deep-sea fishing	7	38.1	266.7
Aquaculture (preferably shrimp farming)	4	32.1	128.4
Cow rearing	4	6.2	24.8
Goat rearing	4	8.3	33.2
Pig rearing	2	1	2
Poultry	3	5.2	15.6

(continued)

Table 2.21 (continued)

	Daily labourers (respondent type 5)		
	LPR	% of vote	LPS5
Agriculture	5	20.4	102
Pisciculture and deep-sea fishing	6	36.4	218.4
Aquaculture (preferably shrimp farming)	6	30.6	183.6
Cow rearing	3	4.1	12.3
Goat rearing	4	5	20
Pig rearing	1	1.1	1.1
Poultry	3	2.4	7.2

Table 2.22 Major Livelihood Preference (LP) of respondents in central Indian Sundarbans

Livelihood	Policy formulator (respondent type 1)		
	LPR	% of vote	LPS1
Agriculture	4	10.6	42.4
Pisciculture and deep-sea fishing	6	25.5	153
Aquaculture (preferably shrimp farming)	7	36.5	255.5
Cow rearing	3	5.6	16.8
Goat rearing	4	4.4	17.6
Pig rearing	1	1	1
Poultry	5	16.4	82
	Agriculturist (respondent type 2)		
	LPR	% of vote	LPS2
Agriculture	2	10.6	21.2
Pisciculture and deep-sea fishing	6	30.2	181.2
Aquaculture (preferably shrimp farming)	7	33.5	234.5
Cow rearing	3	2.6	7.8
Goat rearing	4	3.4	13.6
Pig rearing	1	1	1
Poultry	6	18.7	112.2
	Fisherman (respondent type 3)		
	LPR	% of vote	LPS3
Agriculture	4	9.6	38.4
Pisciculture and deep-sea fishing	6	27.3	163.8
Aquaculture (preferably shrimp farming)	7	36.1	252.7
Cow rearing	5	2.3	11.5
Goat rearing	3	5.5	16.5

(continued)

Table 2.22 (continued)

	Fisherman (respondent type 3)		
	LPR	% of vote	LPS3
Pig rearing	1	1	1
Poultry	4	18.2	72.8
	Livestock owner (respondent type 4)		
	LPR	% of vote	LPS4
Agriculture	4	10	40
Pisciculture and deep-sea fishing	6	28.1	168.6
Aquaculture (preferably shrimp farming)	7	37.8	264.6
Cow rearing	4	4.9	19.6
Goat rearing	3	8.7	26.1
Pig rearing	1	1.1	1.1
Poultry	4	19.4	77.6
	Daily labourers (respondent type 5)		
	LPR	% of vote	LPS5
Agriculture	5	15.4	77
Pisciculture and deep-sea fishing	6	31.2	187.2
Aquaculture (preferably shrimp farming)	7	37.1	259.7
Cow rearing	3	3.6	10.8
Goat rearing	4	5.9	23.6
Pig rearing	1	1	1
Poultry	5	20.3	101.5

the scores for each LP category was given corresponding weights ranging between 1 to 7 Livelihood preference Rank (LPR) and multiplied by the percentage of votes for that option to generate a total weighted score for that LP, which is referred to as Livelihood Preference Score (LPS).

In the final stage, Composite Livelihood Preference Scale (CLPS) was constructed based on Livelihood Preference Scale (LPS) computed as per the expression:

$$\mathrm{CLPS} = \mathrm{LPS1} + \mathrm{LPS2} + \mathrm{LPS3} + \mathrm{LPS4} + \mathrm{LPS5}$$

where, LPS = Livelihood preference Rank (LPR) × % of Vote.

It is to be noted in this context that the sample size of different respondents is variable e.g., the sample size for policy formulator is not like daily labor or fisherman. For policy formulator, we could involve 105 members, but for other respondents n = 390.

Considering the Composite Livelihood Preference Scale (CLPS), the trend of livelihoods in western Indian Sundarbans is Pisciculture and Deep-Sea Fishing (1204.5) > Aquaculture (815.3) > Agriculture (432.5) > Goat Rearing (111.8) > Cow Rearing (77.3) > Poultry (49.5) > Pig Rearing (6.6).

The livelihood preference of the respondents in central Indian Sundarbans is different from those in the western sector. The physico-chemical parameters (preferably the salinity) regulate the livelihood pattern in this sector. Owing to high salinity in this sector, paddy cultivation and agricultural activity did not flourish much, instead shrimp culture in the brackish water canals and ponds have been adopted by majority of the inhabitants. In context to Composite Livelihood Preference Scale (CLPS), the trend of livelihoods in central Indian Sundarbans is Aquaculture, preferably the shrimp culture (1267.0) > Pisciculture and Deep-Sea Fishing (853.8) > Poultry (446.1) > Agriculture (219.0) > Goat Rearing (97.4) > Cow Rearing (66.5) > Pig Rearing (5.1).

We have carried out a survey on the monthly economic profile in the Kakdwip area (21°52′26.50″N and 88°08′04.48″E) of western Indian Sundarbans during 2022 involving stakeholders like agriculturists (n = 512), aquaculturists (n = 295 who are all shrimp culturists), Fisherman (n = 109), livestock owners (n = 215) and service holders (n = 305). The service holders include mostly school teachers, shop owners, daily labors and boat owners. We observed that in this part of Indian Sundarbans, people get maximum income from agriculture and aquaculture (Table 2.23). This may not be a representative picture of central Indian Sundarbans as the region is surrounded by Matla estuary with high saline water.

We observed that most of the people of central Indian Sundarbans are engaged in shrimp culture and many island dwellers practice freshwater prawn (*Macrobrachium rosenbergii*) culture (Fig. 2.37) in their household ponds (that holds the rain water), which fetches considerable income to their family.

Crab fattening also adds substantial income to the aquaculturists of Indian Sundarbans. The juvenile crabs are caught from the wild and their biomass is increased by applying specially formulated protein-rich feed. *Scylla serrata*, popularly known as mud crab is commonly used for fattening purpose (Fig. 2.38). The crabs produced through culture and subsequent fattening is usually sold as exportable products.

The present chapter on traditional livelihood in Indian Sundarbans reveal that there is a great opportunity in the fishery sector and shrimp farming. Improved technology has also boost up the fishery sector in the region. With satellite-based approach, deep sea fishing has also become more profitable. However, natural factors like climate change and human intervention act as guard wall to scale up the opportunity level. To sum up the entire scenario, we carried out a SWOT analysis to explore the traditional livelihood sector in details (Table 2.24).

Table 2.23 Range of monthly economic return (in INR) of the Indian Sundarbans

Category	Agriculture (INR)	Aquaculture (INR)	Fishery and deep-sea fishing (INR)	Livestock rearing (INR)
Agriculturist	8000–15,000	3500–6000	2500–3000	2000–4500
Aquaculturist	2000–3800	6000–10,000	3775–5000	1500–2800
Fisherman	1500–2500	3500–4200	4500–6000	1200–2500
Livestock owners	5000–7500	3800–5000	1500–3000	8000–15,000
Service holders	5000–8000	2000–4500	3000–5000	1500–3000

Fig. 2.37 *Macrobrachium rosenbergii* is cultured in most of the freshwater ponds in Indian Sundarbans

Fig. 2.38 *Scylla serrata* is used for fattening, which is a profitable livelihood for Sundarban people

Table 2.24 SWOT analysis on traditional livelihood sector in Indian Sundarbans

Strength	Weakness	Opportunity	Threat
1] Land and water availability in western sector of Indian Sundarbans; the scenario is not similar in central Indian Sundarbans 2] Wide spectrum of biodiversity including microbial diversity that boost up agricultural productivity 3] Strong knowledge base of the local agriculturist to grow crops in saline soil 4] Developed infrastructures in fishery, tourism, fish landing, and pilgrim related visits 5] Grants and aids from Government agencies and NGO's	1] Hypersalinity of the ambient environment particularly in the central sector of Indian Sundarbans 2] Inadequate supply of water to agricultural fields through irrigation 3] Poor water quality management and disease control practice in the domain of shrimp farming 4] Poor transportation system, that mostly depends on the tidal phase 5] Disaster management is not up to the mark 6] Poor concept development on mangrove-based alternative livelihood	1] Upgradation of local economic profile in the sectors of fishery and animal husbandry 2] Food and nutritional security through apiary, and non-timber forest products 3] Scope of employment generation in the sectors of fishery, apiary, and animal husbandry 4] Scope for introducing agricultural entrepreneurship and climate resilient agriculture practice like salinity resistant crop production 5] Scope for alternative livelihood like oyster culture, halophyte farming, seaweed culture *etc.*	1] Frequent natural disasters in the form of cyclonic storms 2] Tidal surges and damage of embankments 3] Erosion of embankments 4] Lack of skilled labour 5] High population density 6] Poor water quality due to release untreated wastes from shrimp farms, fish landing stations, pilgrims, tourism units, fishing vessels and trawlers *etc.*

Index:

LOW (Excluding fishery sector) | **MEDIUM** (Including fishery sector) | **MAXIMUM** (Excluding fishery sector)

2.4 Take Home Messages

A. Agriculture and fishing activities are the two major pillars of livelihood in Indian Sundarbans although animal husbandry occupies a considerable share in this domain. These pillars are primarily regulated by climatic conditions at local scale. Due to cyclonic depressions coupled with sea level rise, intrusion of saline water takes place in the coastal villages and islands surrounded by saline and brackish water. Even the embankments cannot prevent the entry of saline water from the off-shore region into the brackish and freshwater ecosystem of the islands. In most of the cases, the embankment gets damaged and eroded due to which the intrusion of saline water cannot be prevented. This poses severe adverse impacts on major crops like paddy and vegetables. The major paddy variety cultivated in the islands of Indian Sundarbans is Aman (Kharif). The production of this crop is a function of monsoonal pattern in the area, which has become highly irregular in the lower Gangetic delta complex. Apart from paddy and different types of vegetables, cultivation of betel vine fetches considerable economic return to the people of Indian Sundarbans. The betel leaves locally known as Paan has great demand in several Asian countries like Thailand, Malaysia, Indonesia, Myanmar, Pakistan and Bangladesh. Betel vines are grown in closed bamboo made structure, which are locally known as "Baroj". These are built to protect the plant from the scorchy heat of sun in premonsoon and cold wind in postmonsoon. However, cyclonic depressions often destroy these structures making the cultivation highly vulnerable to natural disaster. In central Indian Sundarbans, which is a high saline belt people are more inclined towards shrimp culture and fishery (mostly polyculture).

B. People of Indian Sundarbans also earn money by rearing cow, goat, sheep, and pig. In recent times, white pigs are reared for pork, which provides a stable income to the people of Indian Sundarbans. The rearing of this species has been given thrust mainly because of their growth in biomass and high rate of breeding frequency; pigs can breed frequently and give birth to around eight to ten piglets during each cycle. The region is blessed with the presence of a valuable genetically rich sheep and goat known as "Garole Sheep" and "Black Bengal Goat" respectively. The livestock sector of Indian Sundarbans produces several marketable items like meat, milk, fiber, skin etc.

C. Considering the Composite Livelihood Preference Scale (CLPS), the trend of livelihoods in western Indian Sundarbans is Pisciculture and Deep-Sea Fishing (1204.5) > Aquaculture (815.3) > Agriculture (432.5) > Goat Rearing (111.8) > Cow Rearing (77.3) > Poultry (49.5) > Pig Rearing (6.6). In this sector, pisciculture primarily includes polyculture practices, where carps, prawns/shrimps are cultured in different combinations. The livelihood preference of the respondents in central Indian Sundarbans is different from those in the western sector. The physico-chemical parameters (preferably the salinity) regulate the livelihood pattern in this sector. Owing to high salinity in this sector, paddy cultivation and agricultural activity did not flourish much, instead shrimp culture in the brackish water canals and ponds have been adopted by majority of the inhabitants. In context to Composite Livelihood Preference Scale (CLPS), the trend of livelihoods in central Indian Sundarbans is Aquaculture, preferably the shrimp culture (1267.0) > Pisciculture and Deep-Sea Fishing (853.8) > Poultry (446.1) > Agriculture (219.0) > Goat Rearing (97.4) > Cow Rearing (66.5) > Pig Rearing (5.1).

D. Crab fattening also adds substantial income to the aquaculturists of Indian Sundarbans. The juvenile crabs are caught from the wild and their biomass is increased by applying specially formulated feed. *Scylla serrata*, popularly known as mud crab is commonly used for fattening purpose. The crabs produced through culture and subsequent fattening is usually sold as exportable products. The island dwellers of Sundarbans are mostly scared to take new ventures in the livelihood sector as there exists uncertainty of markets for oysters, seaweeds, or any mangrove-based snacks that was carried in pilot scale in few pockets of Indian Sundarbans by researchers and NGOs in collaboration with academic institutes.

Annexure 2.1: Feedback Questionnaire

Questions on Provisioning Services of Sundarban Livestock

1. **State your name and profession**

2. **How do you meet up your daily needs?**

3. **What are the common traditional livelihoods in your area?**

4. **In which sector of Indian Sundarbans, you stay? (Tick on appropriate box)**

Tick all that apply.

- ☐ **Western sector**
- ☐ **Central sector**
- ☐ **Eastern sector**

5. **How do you find the role of fishery in upgrading the livelihood of your area?**

Mark only one oval.

1	2	3	4
◯	◯	◯	◯
Least role	**Moderate role**	**High role**	**Highest role**

eeee

6. **How do you score the role of agriculture in boosting the economic profile of your area?**

Mark only one oval.

	1	2	3	4	
Least role	◯	◯	◯	◯	Highest role

Moderate role High role

eeee

7. **How livestock rearing plays the role of economic upgradation in your area?**

Mark only one oval.

	1	2	3	4	
Least role	◯	◯	◯	◯	Highest role

Moderate role High role

eeee

8. **Is there any contribution/role of poultry farming in upgrading the economic profile in your area?**

Mark only one oval.

	1	2	3	4	
Least role	◯	◯	◯	◯	Highest role

Moderate role High role

eeee

9. **How natural disaster like cyclone adversely affect the livelihood?**

Mark only one oval.

	1	2	3	4	
Least effect	◯	◯	◯	◯	Maximum effect

Moderate effect High effect

eeee

10. Is there any effect of aquatic salinity on fishery sector?

Mark only one oval.

	1	2	3	4	
Least effect	◯	◯	◯	◯	Maximum effect

Moderate effect High effect

eeee

11. How do you find the role of Socio-economic profile in the livelihood sector?

Mark only one oval.

	1	2	3	4	
Least role	◯	◯	◯	◯	Highest role

Moderate role High role

eeee

12. What are the major livestock products in your region? Is there any specific trend of production?

____

Google Forms

References

Banerjee K, Vyas P, Chowdhury R, Mallik A, Mitra A (2010) The affects of salinity on the mangrove growth in the lower Gangetic delta. J Ind Ocean Stud 18(3):389–397

Banerjee K, Sengupta K, Raha AK, Mitra A (2013) Salinity based allometric equations for biomass estimation of Sundarban mangroves. Biomass & Bioenergy, (Elsevier) 56:382–391

Bhattacharyya SB, Panigrahi A, Mitra A, Mukherjee J (2010) Effect of physico-chemical variables on the growth and condition index of rock oyster, Saccostrea cucullata (Born) in the Sundarbans India. Indian J Fish 57(3):13–17

Biswas S, Pal N, Zaman S, Mitra A (2019) Samosa and Kachuri from Mangrove associate species: an innovative utilization of coastal flora. Int J Res Appl Sci Eng Tech 7(1):73–79

Biswas S, Zaman S, Mitra A (2020) Development of food products from mangrove associate species: a step towards health security. Our Heritage 68(8):69–77
Chand BK, Trivedi RK, Biswas A, Dubey SK, Beg MM (2012) Study on impact of saline water inundation on freshwater aquaculture in Sundarban using risk analysis tools. Explor Anim Med Res 2(II):170–178
Chowdhury A, Naz A, Bhattacharyya S, Sanyal P (2021) Dynamics of salinity intrusion in the surface and ground water of Sundarban Biosphere Reserve, India. IOP Conf Series: Earth Environ Sci 944:012061. https://doi.org/10.1088/1755-1315/944/1/012061
Dhar S (2011) Impact of climate change on the salinity situation of the Piyali River, Sundarbans, India. J Water Res Prot 3:495–503. https://doi.org/10.4236/jwarp.2011.37059
Dhar I, Sengupta G, Biswas S, Sinha M, Mitra A (2021) Salinity: a major environmental factor in sustainability of the Blue Carbon. J Mech Continua Math Sci 16(11):34–42. https://doi.org/10.26782/jmcms.2021.11.00004
Ghosh R, Banerjee K, Mitra A (2011) Eco-biochemical studies of common seaweeds in the lower Gangetic delta. In: Handbook of marine macroalgae: biotechnology and applied phycology (chapter 3). John Wiley and Sons. Ltd
GoI (2005) Mid term appraisal of the tenth five year plan (2002–2007). Planning Commission, Government of India
GoI (2008) National Livestock Policy 2008. Department of Animal Husbandry, Dairying and Fisheries, Ministry of Agriculture, Government of India
Guha T, Mitra A (2020) Salinity—a crucial factor in ecological sustainability for Sundarbans Mangrove ecosystem. In: Mitra A, Calma MM, Chakrabarty SP (ed) Proceedings: natural resources and their ecosystem services. HSRA Publication, pp 27–35. ISBN 978-81-947216-7-3
Handbook on Fisheries Statistics: 2020, published by Department of Fisheries, Ministry of Fisheries, Animal Husbandry & Dairying, Government of India, New Delhi, pp 196 https://dof.gov.in/sites/default/files/2021-02/Final_Book.pdf
Mitra A, Das KL, Choudhury A (2000b) Oyster farming—a unique approach to utilize our untapped marine resources. J Indian Ocean Stud 8(1&2):103–109
Mitra A, Gangopadhyay A, Dube A, Schmidt ACK, Banerjee K (2009) Observed changes in water mass properties in the Indian Sundarbans (Northwestern Bay of Bengal) during 1980–2007. Curr Sci 97(10):1445–1452
Mitra A, Chowdhury R, Sengupta K, Banerjee K (2010) Impact of salinity on mangroves of Indian Sundarbans. J Coast Environ 1(1):71–82
Mitra A, Saha A, Fernandes M, Pramanick P, Zaman S, Sengupta G (2019) Will non-conventional alternative livelihood scheme work in the framework of mangrove dominated Indian Sundarbans. Ann Mar Sci 3(1):13–17
Mitra A (2013) Sensitivity of mangrove ecosystem to changing climate, vol XIX. Springer, New Delhi, Heidelberg, New York, Dordrecht, London, p 323. ISBN-10: 8132215087; ISBN-13: 978-8132215080. ISBN 978-81-322-1509-7. https://doi.org/10.1007/978-81-322-1509-7
Mitra A (2018) Can species serve as proxy to climate change induced salinity alteration? J Mar Bio Aqua 1–3
Mitra A, Zaman S (2015) Blue carbon reservoir of the blue planet, vol XII. Springer, p 299. ISBN 978-81-322-2106-7. https://doi.org/10.1007/978-81-322-2107-4
Mitra A, Zaman S (2016) Basics of Marine and Estuarine ecology. Springer, India, pp 1–481. ISBN 978-81-322-2707-6
Mitra A, Zaman A (2020) Environmental science—a ground zero observation on the Indian subcontinent, vol XIV. Springer, Chem, p 478. ISBN 978-3-030-49131-4. https://doi.org/10.1007/978-3-030-49131-4
Mitra A, Zaman S (2021) Estuarine acidification: exploring the situation of mangrove dominated Indian Sundarban estuaries, vol XII. Springer, Chem, p 402. ISBN 978-3-030-84792-0. https://doi.org/10.1007/978-3-030-84792-0
Mitra A, Bhattacharyya DP, Choudhury A (2000a) Oyster culture—a solution to algal bloom around shrimp culture farm. Seminar on Mangrove Macrobenthos, Mombasa, Kenya, p 60

Mitra A, Banerjee K, Sengupta K (2011) Impact of AILA, a tropical cyclone on salinity, pH and dissolved oxygen of an aquatic sub-system of Indian Sundarbans. Natl Acad Sci Lett Ind 81(Part II):198–205

Mitra A, Dutta J, Mitra A, Thakur T (2020) Amphan supercyclone: a death knell for Indian Sundarbans. eJ Appl Forest Ecol 8(1):41–48

Mitra A, Zaman S, Pramanick P (2022) Blue economy in Indian Sundarbans: exploring livelihood opportunities, vol XIV. Springer (e-Book), p 403. ISBN 978-3-031-07908-5. https://doi.org/10.1007/978-3-031-07908-5

Mukherjee M, Roy I, Chatterjee P, Basu S, Mitra A (2007) Sea shell craft industry and possibilities of edible oyster culture in coastal West Bengal. In: Sunderban Wetlands, chapter XVII. Department of Fisheries, Aquaculture, Aquatic Resources and Fishing Harbours, Government of West Bengal, pp 271–284

Narhari D (2009) Housing and management of ducks. In: IV World Waterfowl Conference, Thrissur, India, pp 45–47

Pal N, Zaman S, Mitra A (2017) Variation of survival rate of mangrove flora due to salinity in Indian Sundarbans. J Sci Eng Health Manag 1(4):51–54. Published by Techno India University, West Bengal

Pramanick P, Zaman S, Rudra T, Guha A, Mitra A (2015) Heavy metals in a dominant seaweed species from the islands of Indian Sundarbans. Int J Life Sci Pharma Res 5(2):64–71

Pramanick P, Bera D, Banerjee K, Zaman S, Mitra A (2016) Seasonal variation of proximate composition of common seaweeds in Indian Sundarbans. Int J Life Sci Sci Res 2(5):570–578

Pramanick P, Zaman S, Mitra A (2017) Red Seaweed in Sundarban Mangroves can pave the pathway of alternative livelihood. Sci Fed J Glob Warm 1(1):1–14

Pramanick P, Zaman S, Mitra A (2020) Water quality management through seaweed cultivation. Our Heritage 68(8):60–68

Pramanick P, Zaman S, Biswas P, Mitra A (2021) Application of red seaweed catenella repens in alternative livelihood generation in context to blue economy. In: Blue economy: a road map for the future planet, article no. 3. New Millennium Graphics publication, pp. 120–149. ISBN 978-81-951712-5-5

Sinha M, Mukhopadhyay MK, Mitra PM, Bagchi MM, Karmakar HC (1996) Impact of Farakka Barrage on the hydrology and fishery of Hooghly estuary. Estuaries 19(3):710–722

Trivedi S, Zaman S, Ray Chaudhuri T, Pramanick P, Fazli P, Amin G, Mitra A (2016) Inter-annual variation of salinity in Indian Sundarbans. Ind J Geo-Mar Sci 45(3):410–415

Yadav SK, Majumdar K (2020) Mangrove associated seaweeds in Sundarban biosphere reserve, West Bengal, India. Int J Adv Res Biol Sci 7(12):53–62

Chapter 3
Threats to Livelihood Sectors

Contents

3.1 Natural Threat

The natural resources present in the islands, adjoining mudflats, estuaries, creeks etc. of Indian Sundarbans are the foundation stones of regional livelihoods. These natural assets encompass both endemic floral and faunal resources (Fig. 3.1), which are exposed to several natural threats like temperature rise, sea level rise, natural disaster (like cyclones), erosion, salinization, acidification of estuarine water, wave actions, sudden tidal surges etc.

We carried out a stake holder's analysis based on a Questionnaire (*vide* Annexure 3.1) to rank the various natural threats operating in the region. The entire network of the study consist of three stages namely, (i) identification of respondents (policy maker, researcher, fisherman, agriculturist, and local inhabitant) (ii) identification of major threats (8 in number) and (iii) evaluation of the respondent's response to construct the threats scale through ranking and voting. Although threats can be of various types, our list captures the major natural threats operating in the three sectors of Indian Sundarbans. Threats were ranked in terms of their importance by building a Threat Assessment Matrix (TAM). However, as there is high probability of variation in this ranking with the category of respondents, therefore the views of the respondents were also considered (by inclusion of the % of voting along with their

A. Mitra et al., *Climate Resilient Innovative Livelihoods in Indian Sundarban Delta*,
https://doi.org/10.1007/978-3-031-42633-9_3

Fig. 3.1 Indian Sundarban—a hotspot of natural resources

respective ranking factor) and finally Combined Threat Scale (CTS) was constructed based on Threat Scale (TS) computed as per the expression:

$$\text{CTS} = \text{TS1} + \text{TS2} + \text{TS3} + \text{TS4} + \text{TS5}$$

where, TS = Threat Rank (TR) × % of Vote.

The computation was carried out using PYTHON language (Fig. 3.2).

The respondent analysis conducted in the vertical of natural threats operating in the western sector of Indian Sundarbans (Table 3.1) assigned highest value to natural disaster (1060.6), followed by erosion (557.7), wave action and tidal surges (432.2), siltation (319.4), salinification (234.4), acidification (152.7), temperature rise (124.5) and sea level rise (106.5).

An interesting picture is observed while considering the results of respondent analysis for natural threat in the present study region. In Fig. 3.3, acidification has been given a considerable weightage, but the magnitude of threat becomes extremely recessive when the voice of the reserachers are not considered in the analysis (Fig. 3.4). This is because, the concept of lowering the pH of the estuarine waterbodies (referred to as estuarine acidification) is still in an embryonic stage to maximum stakeholders, and hence both ranking and % of vote given by other stakeholders like policy makers, agriculturists, fisherman and local inhabitant (Table 3.1) were lowest for acidification. However, researchers and ecologists working in the domain of hydrological monitoring of Sundarban estuaries (Chakraborty et al. 2013; Jana et al. 2014; Ray Chaudhuri et al. 2014; Hossain et al. 2015; Agarwal et al. 2019; Roychowdhury et al. 2019; Mitra and Zaman 2021) imparted considerable weightage to acidification due to which the value of this threat spikes up in Fig. 3.3. Basically, the main causes of estuarine acidification in western Indian Sundarbans are related to huge emission

```
import pandas as pd
import matplotlib.pyplot as plt
filename = 'Threats.xlsx'
excel_data = pd.ExcelFile(filename)
sheet_names = excel_data.sheet_names
LS_arr = []
for sheet_name in sheet_names:
    psr = excel_data.parse(sheet_name)['TR']
    vote = excel_data.parse(sheet_name)['% of Vote']
    pss = psr * vote
    print(pss)

    LS_arr.append(pss)
# Combine the results from all sheets
LS_combined = sum(LS_arr)
ps = excel_data.parse(sheet_names[0])['Threats']
ps_arr = ps.tolist()
obj = dict(zip(ps_arr, LS_combined))
sorted_LS = sorted(LS_combined, reverse=True)
sorted_names = [key for key, value in sorted(obj.items(), key=lambda x:
x[1], reverse=True)]
print(sorted_LS)
print(sorted_names)
plt.figure(figsize=(20,12))
plt.bar(sorted_names,sorted_LS)
plt.xlabel('Threats',fontsize = 20)
plt.ylabel('Threat Score',fontsize = 20)
#plt.title('Index of dominance of different ponds',fontsize = 20)
plt.show()
```

Fig. 3.2 Computation of respondent analysis to evaluate the natural threat on the livelihood of Indian Sundarban people using PYTHON (as example, the file name has been given 'Threats'; `Threats.xlsx` and the values of TR and % of Vote have been used to feed the Program). The output revealed the Combined Threat Scale (CTS) for all eight major natural threats separately

from local industries and surrounding areas, brick kilns and operations of fishing vessels, trawlers and passenger vessels that will be discussed in details in the next unit of this chapter.

Similar analysis was carried out in the central sector of Indian Sundarbans, where the repondents put maximum weightage on siltation (Table 3.2). This region is noted for hypersaline environment due to Bidyadhari siltation that dates back to late fifteenth century (Chaudhuri and Choudhury 1994; Mitra 2013). Accumulation of silt in the upstream region does not permit the input of fresh water in the region (Fig. 3.5) that has raised the salinity of estuaries like Matla and made the zone unfit for paddy cultivation.

Table 3.1 Threat type with scaling in western Indian Sundarbans

Threat	Policy maker (Respondent Type 1)		
	TR	% of Vote	TS1
Natural disaster	8	25.1	200.8
Wave action and tidal surges	7	14.7	102.3
Temperature rise	4	7.6	30.4
Sea level rise	4	7.1	28.4
Salinification	6	9.8	58.8
Acidification	2	7.9	15.8
Erosion	7	14.9	104.3
Siltation	5	12.9	64.5
Threat	Researcher (Respondent Type 2)		
	TR	% of Vote	TS2
Natural disaster	7	14.9	104.3
Wave action and tidal surges	5	14.7	73.5
Temperature rise	3	5.6	16.8
Sea level rise	3	7.7	23.1
Salinification	5	8.9	44.5
Acidification	5	24.4	122.0
Erosion	6	12.9	77.4
Siltation	4	10.9	43.6
Threat	Fisherman (Respondent Type 3)		
	TR	% of Vote	TS3
Natural disaster	7	26	182.0
Wave action and tidal surges	8	13.9	111.2
Temperature rise	7	5.5	38.5
Sea level rise	3	8.0	24.0
Salinification	5	10.1	50.5
Acidification	2	5.1	10.2
Erosion	8	19.6	156.8
Siltation	7	11.8	82.6
Threat	Agriculturist (Respondent Type 4)		
	TR	% of Vote	TS4
Natural disaster	8	35.9	287.2
Wave action and tidal surges	6	12.7	76.2
Temperature rise	4	5.2	20.8
Sea level rise	3	6.4	19.2

(continued)

Table 3.1 (continued)

Threat	Agriculturist (Respondent Type 4)		
	TR	% of Vote	TS4
Salinification	5	10.3	51.5
Acidification	1	2.8	2.8
Erosion	8	15.0	120
Siltation	6	11.7	70.2
Threat	Local inhabitant (Respondent Type 5)		
	TR	% of Vote	TS5
Natural disaster	7	40.9	286.3
Wave action and tidal surges	6	11.5	69.0
Temperature rise	3	6	18.0
Sea level rise	2	5.9	11.8
Salinification	3	9.7	29.1
Acidification	1	1.9	1.9
Erosion	8	12.4	99.2
Siltation	5	11.7	58.5

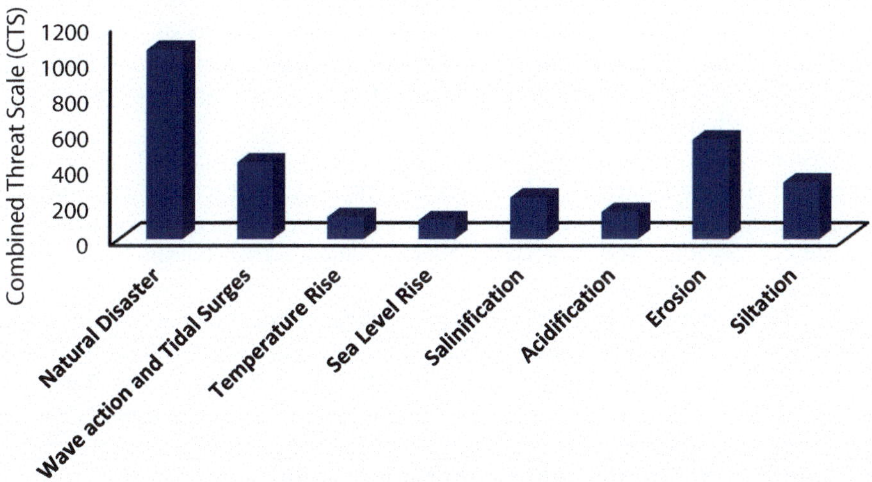

Fig. 3.3 Results of respondent analysis for natural threats in western Indian Sundarbans

The accumulation of silt results in the formation of several new islands locally called 'char' (Fig. 3.6) and thus this deltaic complex is extremely dynamic from the geomorphological point of view.

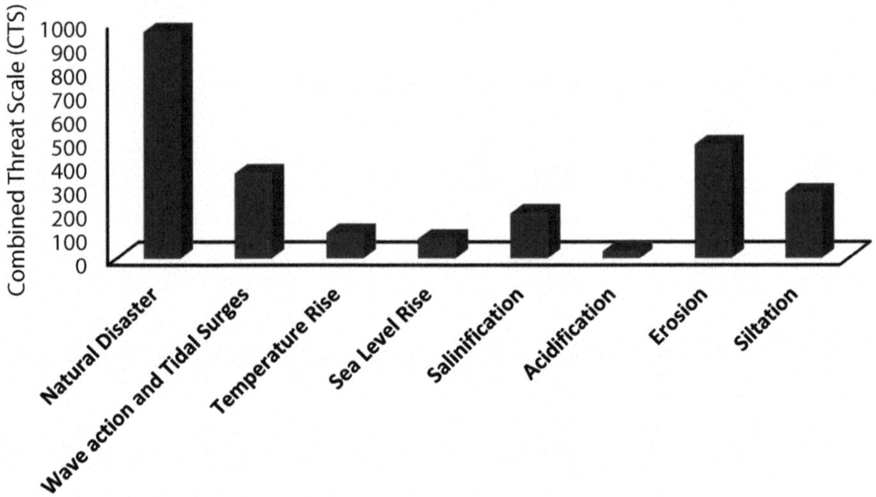

Fig. 3.4 Results of respondent analysis (without considering the feedback of researchers) for natural threats in western Indian Sundarbans

The central sector of Indian Sundarbans is also noted for massive erosion of embankments because of wave actions and tidal surges (Fig. 3.7).

The respondent analysis conducted in the domain of natural threats operating on central sector of Indian Sundarbans assigned highest value to siltation (966.7), followed by erosion (738.8), natural disaster (531.6), salinification (210.2), wave action and tidal surges (175.8), sea level rise (125.5), temperature rise (86.7) and acidification (45.7) (Table 3.2).

To feel the pulse of acidification amongst the respondents, we carried out the analysis considering the inclusion and exclusion of the opinion of researchers, who are important components of the matrix. We observed that for central Indian Sundarbans, the results did not differ and lowest weightage was imparted to estuarine acidification in both cases (Figs. 3.8 and 3.9).

The eastern sector of Indian Sundarbans is adjacent to Bangladesh and is the hotspot of natural disasters. Super Cyclones like Aila, Amphan, Yash crossed over this belt in recent past. This is reflected in the results of respondent analysis (Table 3.3 and Figs. 3.10 and 3.11) where the highest weightage was assigned to natural disaster (969.4), followed by erosion (707.6), wave action and surge (529.5), siltation (353.0), sea level rise (138.5), temperature rise (113.7) and acidification (38.9).

Like central Indian Sundarbans, the respondents in the eastern sector including the researchers did not give weightage to the phenomenon of acidification.

Table 3.2 Threat type with scaling in central Indian Sundarbans

Threat	Policy maker (Respondent Type 1)		
	TR	% of Vote	TS1
Natural disaster	7	17.5	122.5
Wave action and tidal surges	5	9.3	46.5
Temperature rise	4	6.3	25.2
Sea level rise	4	7.9	31.6
Salinification	6	8.8	52.8
Acidification	3	3.6	10.8
Erosion	7	21.4	149.8
Siltation	8	25.2	201.6
Threat	Researcher (Respondent Type 2)		
	TR	% of Vote	TS2
Natural disaster	6	18.9	113.4
Wave action and tidal surges	4	7.6	30.4
Temperature rise	3	5.9	17.7
Sea level rise	3	7.1	21.3
Salinification	5	8.4	42
Acidification	4	3.8	15.2
Erosion	7	22.9	160.3
Siltation	8	25.4	203.2
Threat	Fisherman (Respondent Type 3)		
	TR	% of Vote	TS3
Natural disaster	6	17.1	102.6
Wave action and tidal surges	5	7.2	36
Temperature rise	2	6.6	13.2
Sea level rise	3	8	24
Salinification	4	9.9	39.6
Acidification	1	3.7	3.7
Erosion	7	21.4	149.8
Siltation	7	26.1	182.7
Threat	Agriculturist (Respondent Type 4)		
	TR	% of Vote	TS4
Natural disaster	5	18.1	90.5
Wave action and tidal surges	3	9.1	27.3
Temperature rise	2	6.6	13.2
Sea level rise	3	8.3	24.9

(continued)

Table 3.2 (continued)

Threat	Agriculturist (Respondent Type 4)		
	TR	% of Vote	TS4
Salinification	4	8.7	34.8
Acidification	2	3.7	7.4
Erosion	6	20.7	124.2
Siltation	7	24.8	173.6
Threat	**Local inhabitant (Respondent Type 5)**		
	TR	% of Vote	TS5
Natural disaster	6	17.1	102.6
Wave action and tidal surges	4	8.9	35.6
Temperature rise	3	5.8	17.4
Sea level rise	3	7.9	23.7
Salinification	5	8.2	41
Acidification	2	4.3	8.6
Erosion	7	22.1	154.7
Siltation	8	25.7	205.6

Fig. 3.5 Huge accumulation of silt in and around central Indian Sundarbans

Apart from the respondent analysis, we carried out a ground level survey and based on various literatures available on the issue of impact of natural threats on living resources in the region (that basically form the foundation of livelihood), we conclude a significant spatial variation of natural threats as summarized in Table 3.4.

Fig. 3.6 Newly formed island (locally called 'char') in the midst of the estuary

Fig. 3.7 Erosion of embankments is a common feature in central Indian Sundarbans due to wave action and tidal surges

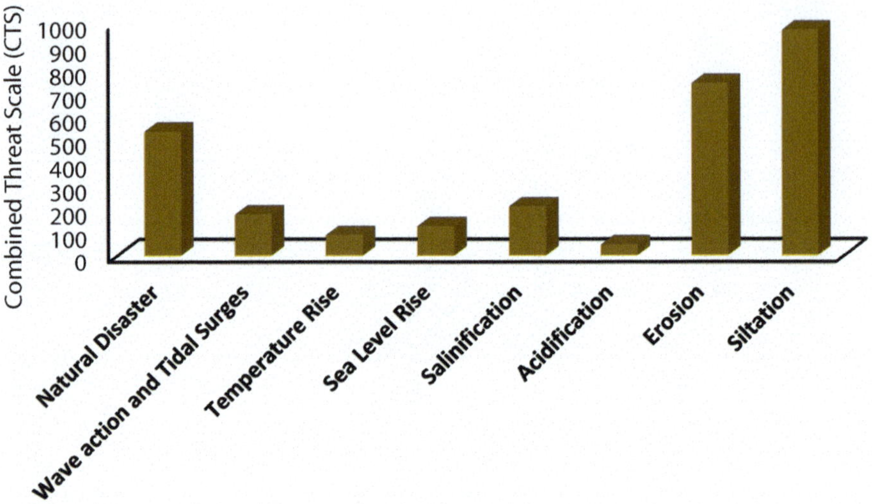

Fig. 3.8 Results of respondent analysis for natural threats in central Indian Sundarbans

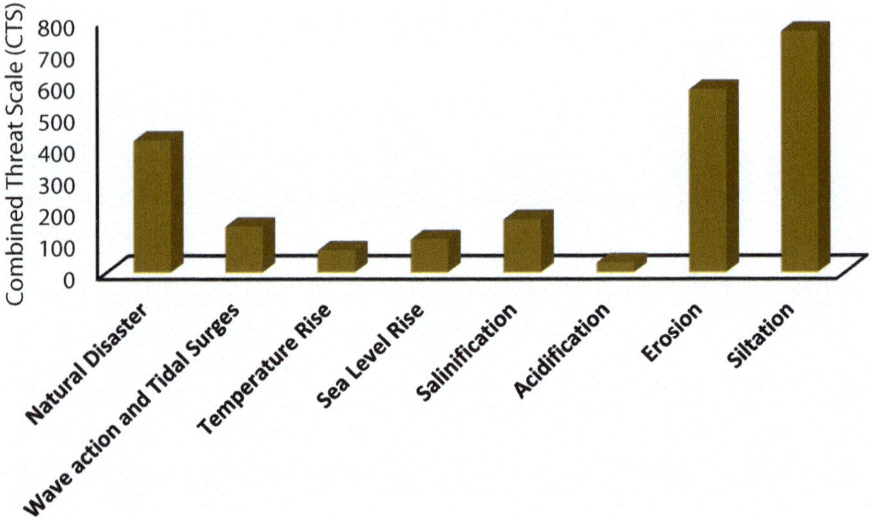

Fig. 3.9 Results of respondent analysis (without considering the feedback of researchers) for natural threats in central Indian Sundarbans

There are several reasons behind the sectoral variation of natural threats in the frame work of Indian Sundarbans.

Table 3.3 Threat type with scaling in eastern Indian Sundarbans

Threat	Policy maker (Respondent Type 1)		
	TR	% of Vote	TS1
Natural disaster	8	22.5	180
Wave action and tidal surges	7	17.6	123.2
Temperature rise	4	7.1	28.4
Sea level rise	4	7.8	31.2
Salinification	3	6.3	18.9
Acidification	2	4.1	8.2
Erosion	7	18.9	132.3
Siltation	5	15.7	78.5
Threat	Researcher (Respondent Type 2)		
	TR	% of Vote	TS2
Natural disaster	7	23.8	166.6
Wave action and tidal surges	5	17.8	89
Temperature rise	3	6.6	19.8
Sea level rise	3	7.1	21.3
Salinification	3	5.9	17.7
Acidification	4	3.7	14.8
Erosion	6	20.9	125.4
Siltation	4	14.2	56.8
Threat	Fisherman (Respondent Type 3)		
	TR	% of Vote	TS3
Natural disaster	8	26.2	209.6
Wave action and tidal surges	6	16.1	96.6
Temperature rise	5	5.1	25.5
Sea level rise	5	7.4	37
Salinification	3	5.3	15.9
Acidification	2	4.3	8.6
Erosion	7	20.9	146.3
Siltation	5	14.7	73.5
Threat	Agriculturist (Respondent Type 4)		
	TR	% of Vote	TS4
Natural disaster	7	28.4	198.8
Wave action and tidal surges	7	16.7	116.9
Temperature rise	4	4.8	19.2
Sea level rise	4	6.4	25.6

(continued)

Table 3.3 (continued)

Threat	Agriculturist (Respondent Type 4)		
	TR	% of Vote	TS4
Salinification	3	4.1	12.3
Acidification	2	2.7	5.4
Erosion	8	23.4	187.2
Siltation	6	13.5	81
Threat	Local inhabitant (Respondent Type 5)		
	TR	% of Vote	TS5
Natural disaster	8	26.8	214.4
Wave action and tidal surges	6	17.3	103.8
Temperature rise	4	5.2	20.8
Sea level rise	3	7.8	23.4
Salinification	2	5.8	11.6
Acidification	1	1.9	1.9
Erosion	6	19.4	116.4
Siltation	4	15.8	63.2

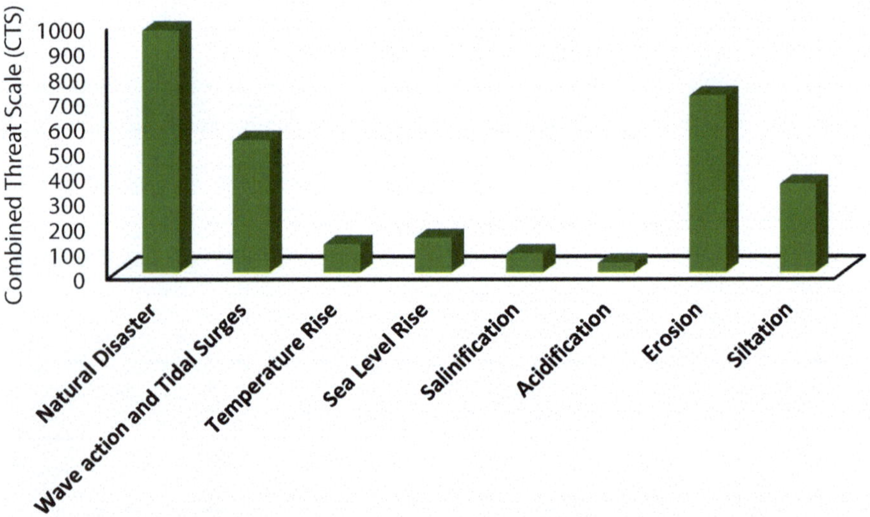

Fig. 3.10 Results of respondent analysis for natural threats in eastern Indian Sundarbans

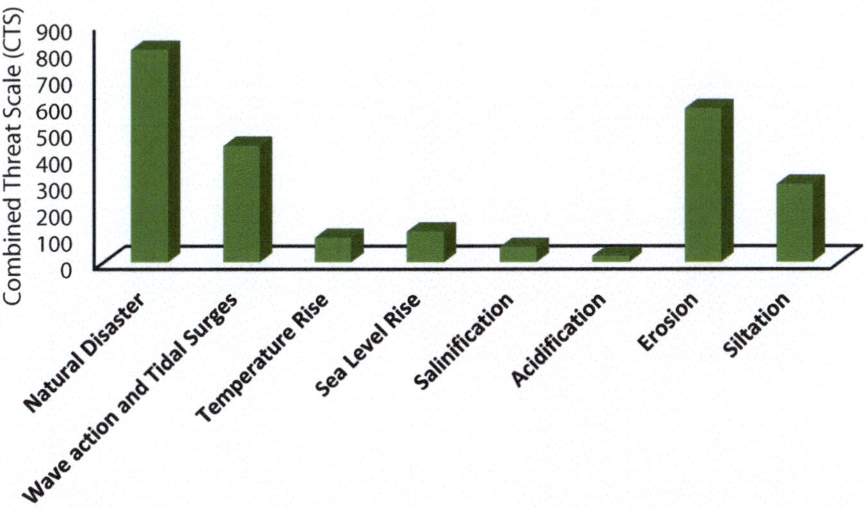

Fig. 3.11 Results of respondent analysis (without considering the feedback of researchers) for natural threats in eastern Indian Sundarbans

The lower Gangetic delta is a cyclone prone zone, where super cyclone around an average speed of 100 ± 10 km hr^{-1} is very common (Table 3.5).

The region also experiences considerable rainfall during July to October. Table 3.6 highlights the average rainfall data for more than three decades (1981–2021).

These natural disasters mostly damage the western and central sectors of Indian Sundarbans due to mass deforestation of mangroves (Fig. 3.12) for developing infrastructures related to fish landing, tourism, jetties, shrimp farms, brick kilns etc., which otherwise could act as a bio-shield to protect the super cyclones.

The eastern sector, on the other hand, is a Reserve Forest (RF) mostly within the core area where the presence of luxuriant mangrove vegetation (Fig. 3.13) not only act as a guard wall against super cyclones, but also arrest the soil particles and prevent erosion.

Temperature rise is relatively high in the western sector owing to presence of industrial belt in the vicinity.

Acidification is an important threat in deltaic Sundarbans, but is more acute in the western sector because of more emission from the ice factories, brick kilns, tourism units, vessels, and trawlers etc. In order to remove the coal tar-based scales for conditioning the fishing vessels and trawlers, appreciable quantum of carbon dioxide is released in the atmosphere which finally leads to acidification through the standardized chemical route as presented in Table 3.7. Conditioning of fishing vessels and trawlers to remove barnacles, oysters etc. is an age-old practice to keep the vessels fit for fishing, which is one of the most important livelihoods of the island dwellers in Indian Sundarbans.

Table 3.4 Impact of natural threat on the resource base in three sectors of Indian Sundarbans

Sector	Western Sector	Central Sector	Eastern Sector	Major Impact
Natural Disaster				1. Uprooting of mangroves 2. Damage to embankments 3. Erosion 4. Intrusion of saline water 5. Damage of agricultural crops 6. Damage of infrastructures 7. Loss of lives and properties 8. Scarcity of fresh water 9. Alteration of fish diversity composition 10. Damage of shrimp farms
Wave action and Tidal Surges				1. Uprooting of mangroves 2. Erosion 3. Intrusion of saline water 4. Damage of embankments 5. Damage of infrastructures like jetties, tourism units, fish landing stations *etc*.
Temperature Rise				1. Salinity alteration 2. Damage of agricultural crops 3. Alteration of fish diversity composition
Sea Level Rise				1. Erosion 2. Damage of embankments 3. Salinification of ground water 4. Damage of agricultural crops 5. Inundation of villages 6. Damage of infrastructures 7. Alteration of fish diversity composition 8. Alteration of hydrological parameters
Salinification				1. Deterioration of ground water quality 2. Damage of agricultural crops 3. Alteration of fish diversity composition (more abundance of trash variety fishes) 4. Alteration of hydrological parameters 5. Stunted growth of mangroves 6. Alteration of mangrove species composition leading to change in mangrove community structure 7. Possibility of extinction of few true mangrove species like *Heritiera fomes*, *Sonneratia* spp. *etc*.

(continued)

Table 3.4 (continued)

Acidification				1. Adverse impact on molluscan community 2. Increase of dissolved heavy metal level due to dissolution of precipitated metallic compounds in the sediment compartment 3. Disruption of the estuarine food chain
Erosion				1. Damage of lives and properties 2. Damage of embankments 3. Intrusion of saltwater in the agricultural fields and fish culture ponds 4. Damage of mangrove vegetation in the inter-tidal mudflats 5. Inundation of villages
Siltation				1. Decreased availability of fresh water 2. Increase of salinity due to decreased dilution 3. Problem in navigation due draft problem, as outcome of siltation

Index:

High Impact	Moderate Impact	Less Impact	Lack of information

3.2 Anthropogenic Threat

The livelihood sector in Indian Sundarbans is exposed to several anthropogenic threats like (i) elevated carbon dioxide level in the atmosphere, (ii) overexploitation of fishes, (iii) acidification of estuarine water, (iv) bioaccumulation of heavy metals in the edible fishes, (v) bioaccumulation of pesticides in the edible food products (preferably fishes), (vi) oil and grease level in the estuarine water, (vii) release of untreated wastes from shrimp farms, and (viii) release of untreated wastes from tourism units. Many of these threats have 'noise' which means there is a mixture of anthropogenic and natural sources, e.g., acidification of estuarine water may occur due to elevated carbon dioxide level in the atmosphere or release of untreated wastes from the adjacent shrimp farms, industries, or fish landing stations.

In this section, all these anthropogenic threats are discussed separately.

Table 3.5 A brief summary of cyclonic storms over the Bay of Bengal region during 1990–2021

S. No.	Year	Name year-wise	Lifetime	MSW (knots)	Duration during ESCS (h)
1	1990$	1990	6 days 6 h	127	60
2	1991$	1991	5 days 12 h	127	36
3	1992#	1992	5 days 12 h	102	48
4	1993#	1993	3 days 6 h	90	3
5	1994$	1994	3 days 3 h	102	18
6	1995#	1995	3 days 21 h	102	15
7	1997$	1997	4 days 6 h	90	12
8	1999#	1999	4 days	90	3
9	1999#	Orissa SuCS	5 days 21 h	140	24
10	2000#	2000	4 days 3 h	102	12
11	2000#	2000	5 days 9 h	90	3
12	2004$	2004	3 days	90	6
13	2006$	Mala	4 days 15 h	100	24
14	2007#	Sidr	4 days 18 h	115	63
15	2008$	Nargis	6 days 6 h	90	9
16	2009$	Aila	2 days	67	24
17	2010#	Giri	2 days 18 h	105	6
18	2013#	Phailin	6 days 3 h	115	54
19	2014#	Hudhud	7 days 6 h	100	27
20	2019	Bulbul	3 days	72	2
21	2019$	Fani	8 days 9 h	116	27
22	2020$	Amphan SuCS	5 days 12 h	102	60
23	2021$	Yash	5 days 4 h	112	12

VSCS very severe cyclonic storm (wind speed between 64 and 89 knots); *ESCS* extremely severe cyclonic storm (wind speed between 90 and 119 knots); *SuCS* super cyclonic storm (wind speed more than 120 knots); $During premonsoon period (March–May); #During monsoon and postmonsoon period (October–December); One knot is equal to 1.8 kmph.
Source Mitra et al. (2022)

3.2.1 *Elevated Carbon Dioxide Level in the Atmosphere*

In Indian Sundarban, burning of fossil fuels like coal or fuel wood for cooking and other household purposes release carbon dioxide in the atmosphere. In addition, industries mainly concentrated at Haldia industrial complex adjacent to western Indian Sundarbans, brick kilns (Fig. 3.14), tourism units, fishing vessels (Fig. 3.15), trawlers, passenger vessels that use diesel release considerable amount of carbon dioxide in the atmosphere.

Table 3.6 Average monthly rainfall data (in mm) collected for Canning region, 24 Parganas, South (West Bengal)

Year	Jan	Feb	Mar	Apr	May	Jun	Jul	Aug	Sep	Oct	Nov	Dec
1981	–	–	–	–	–	–	–	–	–	–	0	0.38
1982	0	1.69	1.65	2.77	0.85	5.69	8.49	10.4	4.23	0.36	0.87	0
1983	0.14	1.67	1.2	1.08	2.51	11.22	6.42	17.79	9.96	3.61	0.25	0.18
1984	0.57	0.03	0	2.29	3.19	23.81	7.26	15.55	6.69	3.28	0	0
1985	0.88	0.21	0.2	1.12	5.11	9.25	9.25	12.2	7.69	3.86	0	0
1986	0.33	0	0.05	1.04	6.2	6.34	10.66	3.35	27.91	8.55	7.54	0.05
1987	0	0.7	0.6	5.8	4.21	4.73	11.55	13.41	9.58	0.81	2	0.32
1988	0	0.92	0.12	0.71	5.49	20.29	17.4	9.28	6.38	4.89	4.66	0
1989	0.07	0.25	0.32	0	5.15	2.83	9.1	10.35	11.34	6.63	0.39	0.08
1990	0	2.48	6.35	3.27	4.79	10.18	16.6	10.77	10.4	5.72	2.41	0.06
1991	2	0.64	0.55	1.13	0.49	16.21	8.64	8.07	8.02	4.56	0.02	0.57
1992	0.92	1.73	0	0.59	6.1	13.31	12.13	13.51	16.1	3.8	0	0
1993	0.03	0	2.75	0.77	4.61	10.34	11.34	11.55	17.73	3.97	0.43	0
1994	0.11	0.84	0.25	3.82	2.35	8.21	14.93	9.12	6.46	2.52	1.92	0
1995	0.23	1.36	0.08	0.61	7.39	11.37	17.6	10	17.84	3.08	6.68	0
1996	0.29	0.3	0.07	0.89	2.56	18.97	11.88	16.72	5.09	9.94	0	0
1997	0.57	0.49	1.36	6.64	3.41	7.12	12.98	15.78	9.36	0.16	0.49	0.48
1998	0.98	0.14	5.97	2.13	5.86	4.94	12.07	12.09	10.41	5.31	5.84	0
1999	0.08	0	0.01	0.05	6.72	8.94	16.47	13.38	14.91	5.96	0.07	0
2000	0.05	1.96	0	2.29	10.79	7.79	15.21	6.34	11.69	4.22	0.06	0.06
2001	0.09	0.01	0.8	3.43	6.8	14.54	10.63	6.87	5.84	5.14	0.62	0
2002	0.85	0	0.53	2.66	4.01	19.81	8.78	10.05	8.31	3.97	4.62	0
2003	0	0.51	0.9	2.26	4.36	10.65	11.9	11.49	9.37	17.09	0.67	1.29
2004	0.15	0.25	0.13	2.26	2.42	8.96	13.35	14.93	7.75	8.77	0	0
2005	1.47	0	4.25	1.24	2.27	8.3	17.89	9.17	11.73	16.74	0	0.03
2006	0	0	0.11	2.06	3.9	4.08	21.16	14.55	17.35	1.045	0.03	0
2007	–	2.8	0.07	1.74	5.63	5.32	20.25	8.97	21.39	3.9	2.13	0
2008	3.46	0.34	0.15	1.4	2.26	12.22	7.16	8.96	9.62	6	–	–
2009	0	0	0.53	0.03	6.52	1.79	13.19	11.47	10.05	4.93	0.34	0
2010	0	0.13	0	0.72	4.25	8.44	8.53	7.56	7.24	4.51	0.12	0.36
2011	0	0.24	1.2	7.31	3.34	14.09	8.98	22.56	7.86	3.1	–	0
2012	1.57	1.01	0.17	3.15	1.52	6.08	10.2	14.78	8.35	4.28	1.86	1.15
2013	0.12	0.18	0.15	0.63	6.36	8.25	8.65	22.08	14.84	9.3	0	0
2014	0	2.33	1.77	0	–	5.78	7.35	–	11.37	1.7	0	0
2015	0.14	0.19	1.83	0.74	4.24	8.21	6.80	13.14	20.14	2.85	0	0
2016	0	0.07	1.09	0	5.10	4.98	5.93	9.68	12.65	3.10	0.12	0

(continued)

Table 3.6 (continued)

Year	Jan	Feb	Mar	Apr	May	Jun	Jul	Aug	Sep	Oct	Nov	Dec
2017	0	0.14	0	0	1.26	11.02	7.80	12.65	9.89	4.02	0	0.14
2018	0.10	1.02	1.21	0.35	3.58	13.65	11.66	14.20	8.42	5.17	0	1.02
2019	1.23	1.45	1.66	0.41	2.97	17.04	13.45	12.06	9.17	8.03	1.05	1.33
2020	1.42	1.27	1.54	0.66	1.56	13.65	12.04	9.28	10.44	11.55	2.40	1.91
2021	1.18	1.34	0.22	0.38	1.11	15.39	19.56	13.76	12.15	14.59	3.26	2.54

Source IMD (Indian Meteorological Department; '–' means data not available; Jan–January; Feb–February; Mar–March; Apr–April; May–May; Jun–June; Jul–July; Aug–August; Sep–September; Oct–October; Nov–November; Dec–December
Source Mitra et al. (2022)

Fig. 3.12 Low population density of mangroves in the western Indian Sundarbans has made the region more vulnerable to cyclone related damage

We estimated the carbon dioxide level at Chemaguri (21° 39′ 58.15″ N and 88° 10′ 07.03″ E) in western Indian Sundarbans from 1984 to 2022 and observed high value of this GHG during the study period (Table 3.8).

Fig. 3.13 Luxuriant mangrove vegetation in the eastern Indian Sundarbans act as a bio-shield against cyclone and wave actions

Table 3.7 Steps of estuarine acidification

Step	Event	Equation
Step 1	Dissolution of atmospheric CO_2 in estuarine water to form aqueous CO_2 along with H_2CO_3	$CO_2 + H_2O \leftrightarrow H_2CO_3$
Step 2	Dissociation of H_2CO_3 to produce HCO_3^- and protons (H^+)	$H_2CO_3 \leftrightarrow HCO_3^- + H^+$
Step 3	Dissociation of HCO_3^- into $CO_3^=$ and protons (H^+), which results in the lowering of pH	$HCO_3^- \leftrightarrow CO_3^= + H^+$

With this data set of more than three decades, we carried out a time series modeling to visualize the trend of the variable using a nonlinear autoregressive model (NAR) by treating the real time carbon dioxide data of 1984–2022 as inputs. The results of this AI based analysis is presented in Table 3.9 and Fig. 3.16.

Massive destruction of halophytes in Sundarbans is an important cause of elevated carbon dioxide in the atmosphere. The shrimp farms, crab farming units and the polyculture practice are mostly carried out by cutting the endemic vegetation, which otherwise could be a potential sink of the gas. In addition, many infrastructures like fish landing stations, tourism units etc. have also expanded at the cost of mangroves.

Fig. 3.14 Brick kilns in Indian Sundarbans is one of the major sources of carbon dioxide

Fishes are important members of both marine and estuarine ecosystems. They are extremely susceptible to a rise in environmental CO_2. Researchers have observed that hypercapnia adversely affects vital physiological functions such as respiration, circulation, and metabolism, and changes in these functions are likely to reduce growth rate and population size through reproduction failure. We carried out a detailed study on the impact of atmospheric carbon dioxide on the Condition Index of *Liza parsia* and *Liza tade* from 2015 to 2023 in the month of April (premonsoon season in the present geographical locale) in every year (Table 3.10).

Condition Index is a function of length and weight of fish species and is useful to evaluate the health of the ambient aquatic phase (Koutrakis and Tsikliras 2003; Froese 2006; Mitra 2013). Condition Index and length–weight relationship studies add beneficial knowledge towards farmed fish producers as these indices are helpful to measure fish growth, population densities, onset of fish maturity, metamorphosis, life history of fishes and overall fish biomass production (Hossain et al. 2006; Araneda et al. 2008; Ferdaushy and Alam 2015).

Fig. 3.15 Fishing vessels in Indian Sundarbans run on diesel, which is one of the major sources of carbon dioxide

Yearly Condition Index (CI) values (from 2015 to 2023) of the cultured species (*Liza parsia* and *Liza tade*) were estimated using the expression (Fulton 1904):

$$K = \frac{\bar{w}}{(TL)^3} \times 10^2$$

where K is the Condition Index, w is the average weight (gm) and TL is the average total length (cm).

We observed two interesting facts from our data sets:

(1) The value of atmospheric carbon dioxide was highest in Kakdwip (in the western Indian Sundarbans) followed by Jharkhali (in the central Indian Sundarbans) and Jhilla (in the eastern Indian Sundarbans), with a sudden dip in 2020 irrespective of all stations, which is an obvious impact of COVID-19 lockdown phase (Mitra 2023).

(2) The condition index values decreased with time in both the species when the atmospheric carbon dioxide showed a rising trend thereby confirming the adverse impact of the gas on the health of the fish species (Figs. 3.17, 3.18 and 3.19).

Table 3.8 CO_2 level (ppm) in the near surface atmosphere at Chemaguri in western Indian Sundarbans (Real time data)

Year	CO_2 level in the near surface atmosphere (in ppm)
1984	359
1985	361
1986	357
1987	360
1988	359
1989	361
1990	362
1991	360
1992	362
1993	364
1994	366
1995	369
1996	366
1997	371
1998	380
1999	378
2000	383
2001	388
2002	390
2003	391
2004	395
2005	397
2006	399
2007	402
2008	401
2009	407
2010	411
2011	413
2012	407
2013	402
2014	399
2015	398
2016	397
2017	402
2018	405
2019	402
2020	346
2021	361
2022	405

Table 3.9 Predicted CO_2 level (ppm) in the near surface atmosphere at Chemaguri in western Indian Sundarbans during 2023–2050

Year	Predicted CO_2 level in the near surface atmosphere (in ppm)
2023	402
2024	384
2025	401
2026	387
2027	404
2028	390
2029	390
2030	395
2031	394
2032	402
2033	401
2034	392
2035	401
2036	396
2037	398
2038	402
2039	404
2040	397
2041	384
2042	407
2043	390
2044	364
2045	407
2046	402
2047	419
2048	478
2049	490
2050	498

3.2.2 *Overexploitation of Fishes*

The estuarine water of Indian Sundarbans sustains a wide range of commercial fishes, mostly consumed by the people of the megacity of Kolkata. The demand is extremely high for fishes and shrimps that are caught from the creeks, inlets, and estuaries of the delta region. Because of this, fisherman of Sundarban want to grab short term profit instead of visualizing the adverse impact on environment.

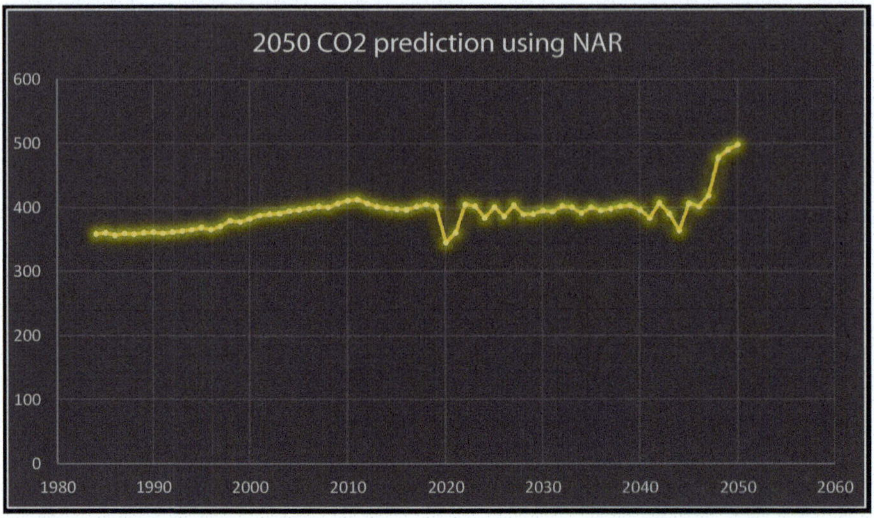

Fig. 3.16 Predicted near surface atmospheric carbon dioxide level for Chemaguri using nonlinear autoregressive neural network model; real-time data from 1984–2022 has been used to train the model

Trawlers with powerful engine have emerged as the biggest threat to fishery sector (Fig. 3.20). A 370-horsepower (HP) engine consumes 40 L of oil in an hour. Besides polluting the aquatic environment, trawlers stay out at sea during the period when the signalling of bans is announced jointly by the Department of Fisheries and the Meteorological Office. This practice has been adopted to accelerate the range of profit caring least for their lives.

In addition to this, excessive use of monofilament nets (Fig. 3.21) to catch surface feeding fishes results in trapping small-sized fishes that cannot attain the adult stage. Intensive misuse of trawl nets for bottom trawling results in the increase of water turbidity, which adversely affects the phytoplankton community that are the primary feed for fishes.

Fishermen of Sundarbans have started indiscriminate commercial exploitation of shrimps of all sizes by modern PVC gill nets instead of traditional cotton nets, for its changeable mesh size, non-biodegradability, more hardness, endurance, and invisible sea water colour. However, this practice have increased the quantum of by-catch and immature fish catch by the fishermen of Sundarbans. Such vulnerable exploitation may lead to an undesirable imbalance in the aquatic food chain of Sundarbans.

Table 3.10 Condition Index (CI) of *Liza parsia* and *Liza tade* in three stations in the western (Kakdwip), central (Jharkhali) and eastern (Jhilla) sectors of Indian Sundarbans along with carbon dioxide level in the near surface atmosphere

Year	*Liza* parsia			*Liza* tade			CO_2 level in the near surface atmosphere (in ppm)		
	Kakdwip	Jharkhali	Jhilla	Kakdwip	Jharkhali	Jhilla	Kakdwip	Jharkhali	Jhilla
2015	0.73 ± 0.03	0.81 ± 0.03	1.02 ± 0.03	0.60 ± 0.03	0.68 ± 0.03	0.67 ± 0.03	365	359	347
2016	0.71 ± 0.03	0.73 ± 0.03	0.95 ± 0.03	0.51 ± 0.03	0.62 ± 0.03	0.64 ± 0.03	374	368	352
2017	0.63 ± 0.03	0.70 ± 0.03	0.91 ± 0.04	0.42 ± 0.03	0.56 ± 0.03	0.61 ± 0.03	389	377	363
2018	0.58 ± 0.04	0.81 ± 0.04	0.88 ± 0.04	0.41 ± 0.03	0.50 ± 0.04	0.59 ± 0.03	397	380	369
2019	0.49 ± 0.04	0.83 ± 0.04	0.82 ± 0.05	0.40 ± 0.04	0.47 ± 0.04	0.56 ± 0.04	405	391	375
2020	1.01 ± 0.05	1.05 ± 0.06	1.07 ± 0.06	1.03 ± 0.04	1.05 ± 0.05	1.08 ± 0.05	353	348	337
2021	0.99 ± 0.07	1.01 ± 0.07	1.04 ± 0.08	1.02 ± 0.07	1.03 ± 0.07	1.05 ± 0.08	390	387	355
2022	0.69 ± 0.05	0.85 ± 0.06	0.99 ± 0.07	0.89 ± 0.08	0.91 ± 0.06	0.96 ± 0.07	413	404	381
2023	0.65 ± 0.06	0.79 ± 0.07	0.93 ± 0.07	0.55 ± 0.08	0.60 ± 0.06	0.72 ± 0.06	427	415	395
r-value	−0.5618	−0.2428	−0.5873	−0.2843	−0.2158	−0.3000	–	–	–
p-value	$p < 0.01$	IS	$P < 0.01$	IS	IS	IS	–	–	–

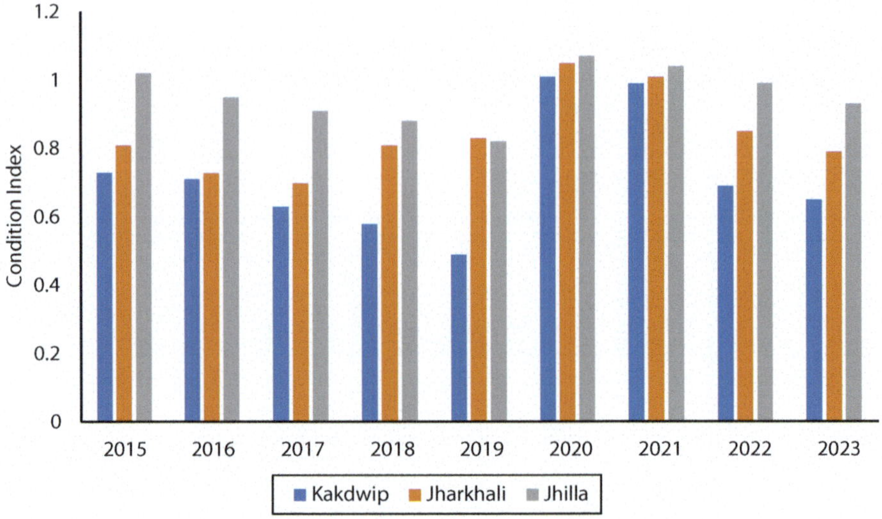

Fig. 3.17 Condition Index value of *Liza parsia* in three selected stations in Indian Sundarbans

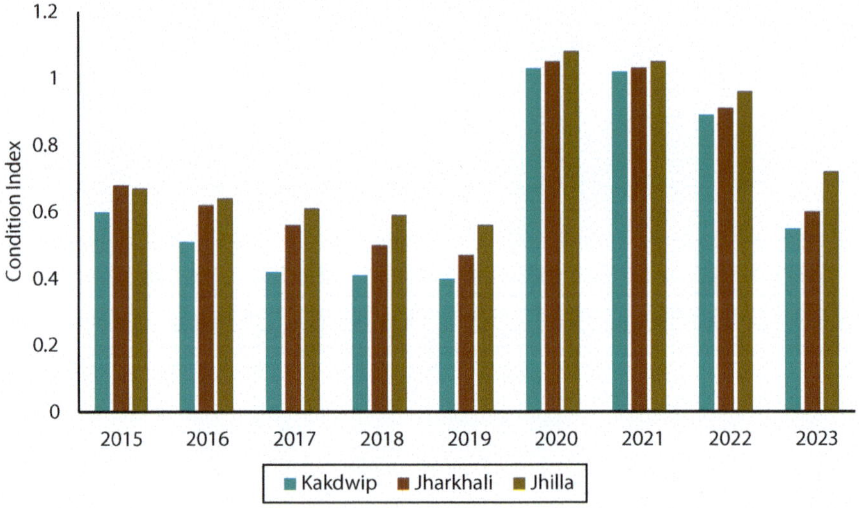

Fig. 3.18 Condition Index value of *Liza tade* in three selected stations in Indian Sundarbans

3.2.3 Acidification of Estuarine Water

Acidification is not an exaggerated future scenario. It is a reality and the pulse has been felt even in the estuaries of Indian Sundarbans (Chakraborty et al. 2013; Ray Chaudhuri et al. 2014; Agarwal et al. 2019; Roychowdhury et al. 2019; Mitra and

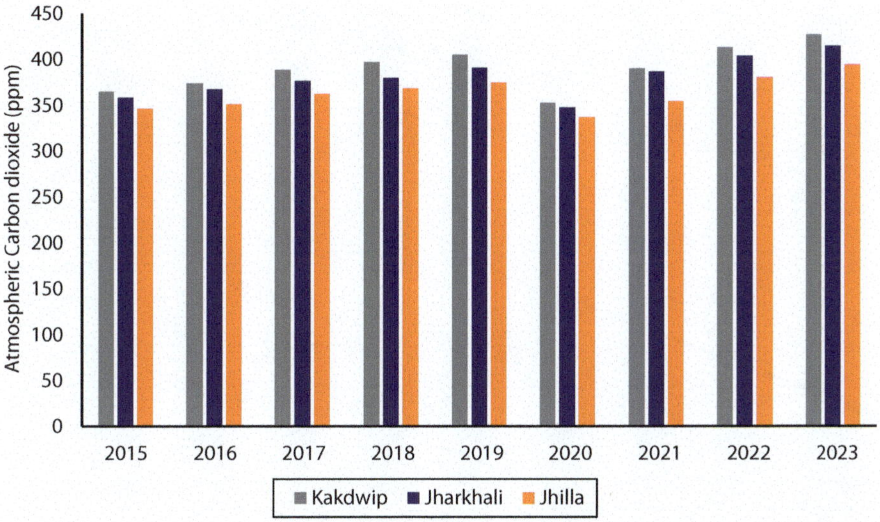

Fig. 3.19 Near surface atmospheric carbon dioxide level in three stations in Indian Sundarbans

Fig. 3.20 Fishing trawlers often disobey the ban period and go for over-exploitation of fishes

Fig. 3.21 Monofilament nets result in trapping immature/juvenile fishes

Zaman 2021). We analyzed the situation in 24 stations of Indian Sundarbans spread across the western, central, and eastern sectors of Indian Sundarbans (Table 3.11).

In few of these stations variety of infrastructures starting from tourism to pilgrims exist (Figs. 3.22 and 3.23).

Acidification can corrode these infrastructures, which is sustaining the livelihoods of a larger section of the people in this region. In addition to these, oysters and various types of molluscs are used in this region as a source of lime (mostly carried out illegally), which will soon see the sign of complete full stop. We could analyze the situation from our 'Oyster Watch Programme' that we carried out in 24 islands of Indian Sundarbans during March, 2023 where we tried to figure out the inter-relationship between oyster shell (Fig. 3.24) weight and ambient aquatic pH (Table 3.11).

We observed significant positive correlation between oyster shell weight and surface water pH (used as proxy to acidification) confirming the dissolution of the shell of the species at lower pH ($r = 0.9691$; $p < 0.01$). Oyster, being the main consumer of phytoplankton may result in phytoplankton bloom and lowering of dissolved oxygen (DO). This situation is extremely adverse for aquatic lives thriving in the system.

Constant flow of water over sediment also makes the underlying substratum acidic. Mangrove or acid sulfate soils are not suitable for shrimp pond culture due to their high organic matter contents and acidic nature that require a high-water exchange rate and low stocking density. A pond built on mangrove soil will also encounter the problems of hydrogen sulfide and ammonia accumulation in the pond bottom.

Table 3.11 Sampling stations in Indian Sundarbans to carry out Oyster Watch Programme

Sl. No.	Sampling station	Latitude and longitude	Oyster shell weight (gm)	Surface water pH
1	Muriganga	21° 38′ 25.86″ N; 88° 08′ 53.55″ E	41.00	8.31
2	Saptamukhi	21° 36′ 02.49″ N; 88° 23′ 47.18″ E	44.56	8.33
3	Thakuran	21° 49′ 43.17″ N; 88° 33′ 21.57″ E	42.56	8.32
4	Herobhanga	21° 59′ 34.32″ N; 88° 41′ 46.52″ E	41.90	8.31
5	Ajmalmari	21° 51′ 34.72″ N; 88° 39′ 00.68″ E	42.09	8.32
6	Dhulibasani	21° 47′ 06.62″ N; 88° 33′ 48.20″ E	44.15	8.34
7	Chulkathi	21° 41′ 53.62″ N; 88° 34′ 10.31″ E	44.44	8.34
8	Arbesi	22° 11′ 43.14″ N; 89° 01′ 09.4″ E	43.78	8.33
9	Jhilla	22° 09′ 51.53″ N; 88° 57′ 57.70″ E	45.41	8.34
10	Pirkhali	22° 06′ 00.97″ N; 88° 51′ 06.40″ E	37.23	8.30
11	Panchmukhani	21° 59′ 41.58″ N; 88° 54′ 14.71″ E	44.78	8.33
12	Harinbhanga	21° 57′ 17.85″ N; 88° 59′ 33.24″ E	34.22	8.29
13	Khatuajhuri	22° 03′ 06.55″ N; 89° 01′ 05.33″ E	42.10	8.32
14	Chamta	21° 53′ 18.56″ N; 88° 57′ 11.40″ E	43.19	8.33
15	Matla	21° 53′ 15.30″ N; 88° 44′ 08.74″ E	32.09	8.26
16	Chandkhali	21° 51′ 13.59″ N; 89° 00′ 44.68″ E	44.90	8.34
17	Goashaba	21° 43′ 5.64″ N; 88° 46′ 41.44″ E	42.29	8.32
18	Gona	21° 41′ 15.44″ N; 88° 54′ 31.09″ E	45.54	8.34
19	Chhotahardi	21° 44′ 42.24″ N; 88° 44′ 17.79″ E	39.89	8.30
20	Baghmara	21° 39′ 04.45″ N; 89° 04′ 40.59″ E	47.05	8.34
21	Mayadwip	21° 35′ 50.23″ N; 88° 47′ 09.95″ E	46.22	8.34
22	Jambu Island	21° 35′ 42.03″ N; 88° 10′ 22.76″ E	31.19	8.25
23	Lothian	21° 38′ 21.20″ N; 88° 20′ 29.32″ E	40.88	8.30
24	Sagar Island	21° 38′ 51.55″ N; 88° 02′ 20.97″ E	29.08	8.22

In the acid sulfate soil areas, the soil will develop high acidity when dried and then flooded, which will lead to difficulty in stabilizing the pH of the pond water and in inducing the growth of plankton during the culture period. This will severely affect the production of shrimp posing a negative effect on the Return on Investment (ROI).

3.2.4 Bioaccumulation of Heavy Metals in the Edible Fishes

In Indian Sundarbans, this threat is not visible in a greater scale, but because of the release of wastes from factories and industries adjacent to the World Heritage Site, effluents contaminate the estuarine water with wastes of all categories. Heavy metals, being an important component of these wastes are non-biodegradable in nature and

Fig. 3.22 Jetty pillars are the attachment sites of oysters which are corroding due to acidification

Fig. 3.23 Kapil Muni temple at Sagar Island is the livelihood of many local people, which might face danger due to approaching sea with low water pH

Fig. 3.24 *Saccostrea cucullata* (common mangrove oyster), the candidate species in the Oyster Watch Programme

get accumulated within the fish muscles. Many of these species have high export of values, but unfortunately, they are rejected from the export basket due to deterioration of quality.

A very recent studies conducted by us on March, 2023 on the bioaccumulation pattern of heavy metals in the fish sampled from Kakdwip (21° 52′ 26.50″ N and 88° 08′ 04.48″ E), and Jhilla (22° 09′ 51.53″ N and 88° 57′ 57.07″ E) exhibited high values of Zn and Pb in the fish muscle sampled from Kakdwip region. This is mainly because of extreme anthropogenic stress in the region due to presence of fish landing stations, passenger jetties, frequent use of antifouling paints for conditioning vessels, boats, and trawlers etc. These are the point sources of heavy metals in the estuarine water of Sundarbans. Jhilla forest, on the other hand is in the extreme east of Indian Sundarbans, where the anthropogenic foot steps are minimum (Fig. 3.25). People hardly go there in the dense mangrove forest and this practically area of zero human interferences due which the fishes exhibit very minimum bioaccumulation (Table 3.12).

Table 3.12 Bioaccumulation of Zn and Pb (in ppm dry wt.) in the selected fin fish species sampled from the estuarine water adjacent to Kakdwip and Jhilla forest

S. No.	Commercially important fin fish	Kakdwip in Western Indian Sundarbans		Jhilla in Eastern Indian Sundarbans	
		Zn (ppm)	Pb (ppm)	Zn (ppm)	Pb (ppm)
1	 *Tenualosa ilisha* (Family: Clupeidae)	198. 88	33.65	76.22	BDL
2	 *Pama pama* (Family: Sciaenidae)	154.15	12.65	41.08	2.30
3	 *Pampus* spp. (Family: Stromateidae)	249.56	6.04	103.36	1.77
4	 *Ilisha elongata* (Family: Pristigasteridae)	332.10	38.17	64.07	1.05
5	 *Lates calcarifer* (Family: Centropomidae)	448.15	29.34	74.65	1.11

(continued)

Table 3.12 (continued)

S. No.	Commercially important fin fish	Kakdwip in Western Indian Sundarbans		Jhilla in Eastern Indian Sundarbans	
		Zn (ppm)	Pb (ppm)	Zn (ppm)	Pb (ppm)
6	 *Pangasius pangasius* (Family: Pangasiidae)	541.24	17.21	29.67	BDL
7	 *Liza parsia* (Family: Mugilidae)	456.44	39.20	104.36	2.95
8	 *Liza tade* (Family: Mugilidae)	502.78	12.66	78.56	3.85
9	 *Tenualosa toli* (Family: Clupeidae)	511.98	17.02	66.11	BDL
10	 *Polynemus paradiseus* (Family: Polynemidae)	100.43	2.11	47.87	BDL
11	 *Otolithoides biauritus* (Family: Sciaenidae)	99.56	1.93	NS	NS

(continued)

Table 3.12 (continued)

S. No.	Commercially important fin fish	Kakdwip in Western Indian Sundarbans		Jhilla in Eastern Indian Sundarbans	
		Zn (ppm)	Pb (ppm)	Zn (ppm)	Pb (ppm)
12	 *Tachysurus jella* (Family: Ariidae)	432.90	34.89	34.90	2.56
13	 *Sciaena biauritus* (Family: Sciaenidae)	257.99	32.88	56.09	1.62
14	 *Eleutheronema tetradactylum* (Family: Polynemidae)	122.55	1.39	23.90	BDL

BDL means Below Detectable Level; NS means non availability of fish sample

3.2.5 Bioaccumulation of Pesticides in Fishes

The deltaic complex of Indian Sundarbans has seven major estuaries, which sustain a wide variety of finfishes and shell fishes. However, these estuaries receive several categories of industrial and anthropogenic wastes from the city of Kolkata, Howrah and Haldia port region. Several agricultural fields grow paddy and vegetables that exist in the vicinity of Hooghly and Matla estuaries, which drain considerable load of pesticides into the nearby estuaries as run-off. Our sampling site Diamond Harbour (22° 11′ 4.2″ N and 88° 11′ 22.2″ E) is situated along the Hooghly estuaries in the maritime state of West Bengal that receives waste of various categories from the industries and agricultural fields. Although DDT is officially banned in Indian sub-continent, but its illegal use by the agriculturist is continuing uninterruptedly. Apart from this, DDT is also used for disease vector control. These sources drain DDT into the estuaries of Hooghly River and finally bioaccumulated within the body tissues of the resident organisms. The half-life of DDT is quite long and ranges from 2–15 years.

Fig. 3.25 Jhilla forest in eastern Indian Sundarbans is a tiger habitat with almost no anthropogenic footsteps. Nets at the open spaces are provided to keep the tiger within the forest territory

In the present case study carried out during March 2023, tDDT concentration in the fish muscles varies as per the order *Lates calcarifer* (54.07 ± 0.89 ng g^{-1} lipid wt.) > *Penaeus* sp. (47.22 ± 0.99 ng g^{-1} lipid wt.) > *Tenualosa ilisha* (37.83 ± 0.97 ng g^{-1} lipid wt.) > *Pangasius* sp. (30.96 ± 0.92 ng g^{-1} lipid wt.) > *Pampus* sp. (25.18 ± 1.03 ng g^{-1} lipid wt.) > *Polynemus paradiseus* (25.11 ± 0.95 ng g^{-1} lipid wt.) (Table 3.13). Overall, these concentrations are lower than those usually observed in different parts of the World, apparently as a result of high tidal actions at regular intervals that allow dispersion of the pollutants and subsequently reduces the probability of bioaccumulation.

The presence of pesticides in the edible fishes is a matter of great concern as it can reduce the export value of the product. This will disrupt the fishery-based livelihood and economic profile of the region.

3.2.6 Oil and Grease Level in the Estuarine Water

The estuarine water in Indian Sundarbans is saturated with oil and grease that are released from fishing vessels, trawlers, passenger vessels and oil wastes from refineries in Haldia region. Oil and grease from a film on the surface water, which

Table 3.13 List of fish species with respective tDDT levels (ng/gm) in their muscle

Sl. No.	Scientific name	Pictorial representation	tDDT (ng/gm)
1	*Tenualosa ilisha*		37.83 ± 0.97 (15.83–52.77)
2	*Polynemus paradiseus*		25.11 ± 0.95 (11.61–41.57)
3	*Penaeus* sp.		47.22 ± 0.99 (18.63–66.86)
4	*Lates calcarifer*		54.07 ± 0.89 (20.57–70.92)
5	*Pangasius* sp.		30.96 ± 0.92 (13.61–38.50)
6	*Pampus* sp.		25.18 ± 1.03 (8.58–35.63)

does not allow the sunlight and oxygen to diffuse into a greater depth. This results in the death of phytoplankton community subsequently leading to disruption of the entire food-web existing in the region. We carried out a study on the oil and grease levels in and around Namkhana (21° 45′ 53.7″ N; 88° 13′ 51.5″ E) located in western sector of Indian Sundarbans during 2011–2020. In order to assess the magnitude of threat posed by oil and grease on the phytoplankton community, a simultaneous study was carried out to evaluate the diversity of phytoplankton using Shannon-Weiner Species Diversity Index as proxy.

Phytoplankton samples (n = 4) collected from the surface water around Namkhana station were observed with a ZEISS research microscope coupled with an image analysing system. The cell identifications were based on standard taxonomic keys (Botes 2003; Verlencar 2004).

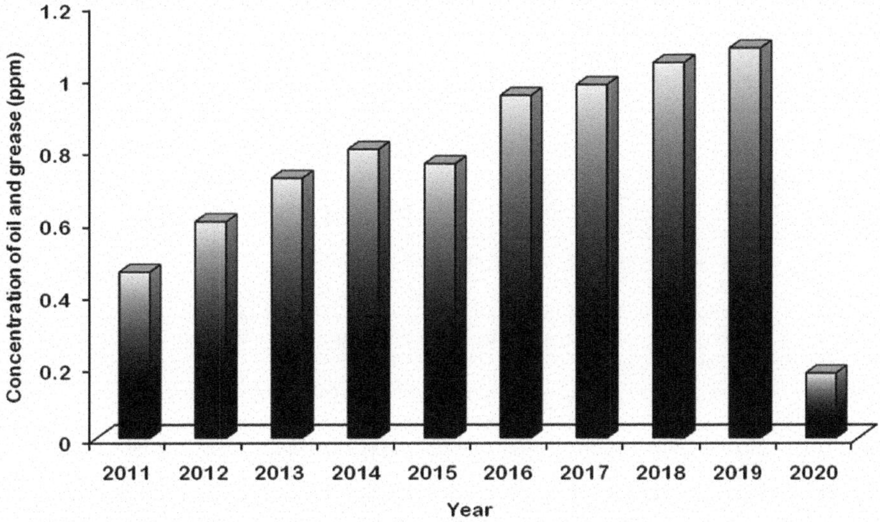

Fig. 3.26 Temporal variations of oil and grease level (ppm)

We observed an increasing trend of oil and grease levels, in the estuarine water during 2011–2019, the period characterized by normal plying of fishing vessels, trawlers, oil tankers etc. to the estuarine water (Fig. 3.26).

There is also a major contribution of oil and grease from the fish-landing stations located at Namkhana and Frasergunge, which are very close to each other. In addition, oil and grease also enter the aquatic system through de-ballasting (which is a normal operation carried out in port area) refineries, municipal and industrial wastes etc.

The Shannon-Weiner Species Diversity Index value computed by Python Programme (Fig. 3.27) exhibited significant temporal variations with decreasing trend during 2011–2019 (Fig. 3.28).

The year 2020 exhibited an opposite picture when oil and grease levels dipped down and diversity index of phytoplankton increased sharply. This is the result of COVID-19 lockdown phase when there was a complete ceasefire of many of the anthropogenic activities. The peak of phytoplankton diversity during 2020 when there was minimum oil and grease in the estuarine water (Fig. 3.28) confirms the fact that oil and grease pose significant adverse impact on the primary and secondary productivity of the estuarine water.

The secondary production, which includes the fishes, is the primary source of livelihood. Thus, any adverse impact on phytoplankton community is likely to affect the fishery-based livelihood of Sundarban people.

```
import pandas as pd
import numpy as np
import matplotlib.pyplot as plt

data1 =  pd.read_excel('phytoplankton_1.xlsx',sheet_name=0)
stock1 = data1['Standing Stock (No. × 105/L)']
p1 = stock1/stock1.sum()
shannon_weiner_index_sampling point1= -np.sum(p1 * np.log(p1))
print('Shannon weiner index for sampling
point 1 = ',shannon_weiner_index_sampling point1)

data2 =  pd.read_excel('phytoplankton_1.xlsx',sheet_name=1)
stock2 = data2['Standing Stock (No. × 105/L)']
p2 = stock2/stock2.sum()
shannon_weiner_index_sampling point2= -np.sum(p2 * np.log(p2))
print('Shannon weiner index for sampling
point 2 = ',shannon_weiner_index_sampling point2)

data3 =  pd.read_excel('phytoplankton_1.xlsx',sheet_name=2)
stock3 = data3['Standing Stock (No. × 105/L)']
p3 = stock3/stock3.sum()
shannon_weiner_index_sampling point3= -np.sum(p3 * np.log(p3))
print('Shannon weiner index for sampling
point 3 = ',shannon_weiner_index_sampling point3)

data4 =  pd.read_excel('phytoplankton_1.xlsx',sheet_name=3)
stock4 = data4['Standing Stock (No. × 105/L)']
p4 = stock4/stock4.sum()
shannon_weiner_index_sampling point4= -np.sum(p4 * np.log(p4))
print('Shannon weiner index for sampling
point 4 = ',shannon_weiner_index_sampling point4)

Station = ['sampling point 1']
SWI = [shannon_weiner_index_sampling
point1, shannon_weiner_index_sampling
point2, shannon_weiner_index_sampling
point3, shannon_weiner_index_sampling point4]
plt.figure(figsize=(20,12))
plt.bar(ponds,SWI)
plt.xlabel('sampling point')
plt.ylabel('Shannon weiner index')
plt.title('Shannon weiner index of different sampling point')
plt.show()
plt.show()
```

Fig. 3.27 Computation of Shannon Weiner Species Diversity Index for phytoplankton using Python

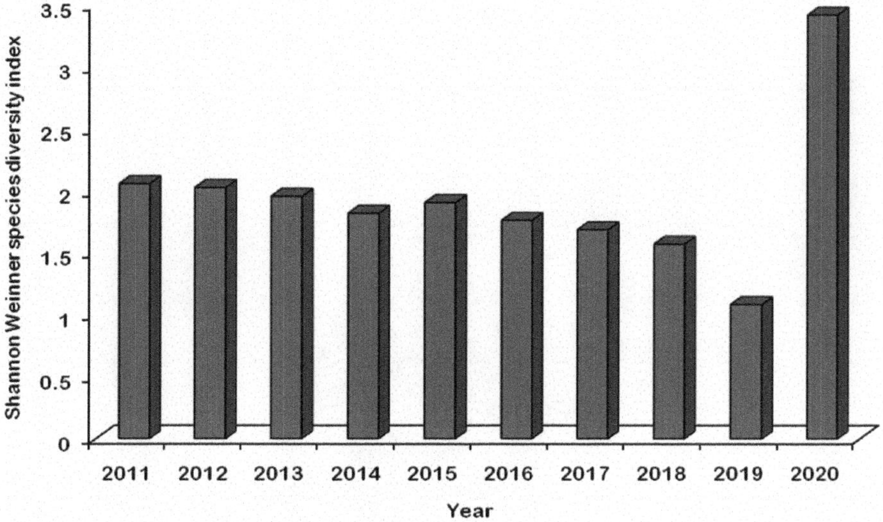

Fig. 3.28 Temporal variations of Shannon Weinner Species Diversity Index of phytoplankton

3.2.7 Release of Untreated Wastes from Shrimp Farms

Shrimp farming is a well-accepted livelihood in the framework of both Indian and Bangladesh Sundarbans. It is carried out by traditional, modified extensive and semi-intensive methods. These methods differ from each other mainly in terms of (i) stocking density of post larval phase of shrimp, (ii) engineering design of the pond, (iii) nature of feeding (whether natural phytoplankton or artificial feed in the form of pellet), (iv) water and soil quality and (v) pond management. It is however, a tragedy that majority of the shrimp farmers do not construct the treatment pond for treating the wastes generated from these farms preferably which are semi-intensive in nature. The basic cause behind such mindset is to achieve a short-term gain at the cost of environment. This causes accumulation of organic load in the pond bottom, increase of ammonia and hydrogen sulfide levels, nutrient overloading in the pond water and depletion of oxygen. All these are detrimental to the growth and survival of shrimps, and the situation is overcome usually through water exchange (Fig. 3.29).

3.2.8 Release of Untreated Wastes from Tourism Units

Indian Sundarban is the hotspot of tourists who may be of several categories like religious tourist, academic tourist, aesthetic tourist, forest lovers etc. Considering their demands, food preference and facilities, several types of tourism units have developed in Indian Sundarbans particularly in the western and central sectors of the region. Beaches like Bakkhali (21° 33′ 33.84″ N and 88° 15′ 57.66″ E), Sagar Island

Fig. 3.29 Water exchange from a semi-intensive shrimp farm in Sundarban

(21° 38′ 51.55″ N and 88° 02′ 20.97″ E) etc. in the western Indian Sundarbans are thickly populated with tourists. Unregulated tourism units have also flourished in the adjacent areas of Indian Sundarbans that release wastes of complex characters (Figs. 3.30 and 3.31).

These tourism units release wastes which mainly consist of nutrients like nitrate and phosphate. Along with these nutrients food wastes and oil are also released into the adjacent water bodies that degrade the quality of the water in terms of survival and growth of the organisms. This is one of the major threats in the livelihood domain of Sundarban people as it leads to (i) depletion of oxygen, (ii) eutrophication and (iii) bloom of harmful algae. All these may cause mortality of the aquatic lives, which has high probability to retard the economic profile of the Sundarban people.

3.3 'Noise' in Threat Level

Researches related to human impacts on climate change and the inter relationships between natural forces and climate change date back to eighteenth century. However, in most of these researches it is difficult to get a clear 'signal' or cause behind a phenomenon. Several factors appear on the plate of the readers as 'noise' and this creates confusion. As stated earlier that climatic variations are often caused by natural factors (Fig. 3.32), but according to Shabecoff (1988) "… the cause of warmer temperatures were the burning of fossil fuels and not solely a result of natural

Fig. 3.30 Tourism units adjacent to Indian Sundarban region

variation …". At this point, the interference of several factors to explain climate change appears as 'noise' and a sense of uncertainty appears on the surface.

In this unit we will focus on three proxies of climate change which are relevant in explaining the footsteps of climate change in the present study area. These are (i) Sea level rise, (ii) warming of estuarine water and (iii) alteration of salinity.

3.3.1 Sea Level Rise (SLR)

The contribution of glacier and ice sheet melting is considered as the major actor for Global Mean Sea Level (GMSL) alteration, although thermal expansion of water cannot be ignored. It has increased from 1.4 mm yr^{-1} over the period 1901–1990 to 2.1 mm yr^{-1} over the period 1970–2015, and 3.2 mm yr^{-1} over the period 1993–2015. During 2006–2015, the value touched 3.6 mm yr^{-1}. Thus, from 1970 onwards, the anthropogenic factor became predominant.

It is a real picture almost uniformly throughout the World. However, for Indian Sundarbans, the value is 3.14 mm/yr (Hazra et al. 2002). Now, why this variation? Some researchers point towards the melting of Himalayan glaciers that brings huge freshwater into the region through several states (Mitra et al. 2009), while some schools point towards the phenomenon of 'subsidence'. The Sundarban delta region

Fig. 3.31 Tourists enjoying the beach at Bakkhali in western Indian Sundarbans

is located on the western part of the Nagalushai Geosyncline, which has a hinge in the basement complex. This hinge separates the stable shelf part and the deeper part of the basin in the east and southeast. Thus, the Bengal Basin has strata thickening eastward towards the Indo-Myanmar ranges. The base of the hinge is located at Sagar Island and the crack extends to Mymensingh and Sylhet (Srihatta) districts of Bangladesh. Hence, the Sundarban delta is tilting gradually towards the east, resulting in submergence of the islands here. It has been found that the rate of subsidence along the hinge zone is less than that in the deeper part of the basin. It is a fact that the sediments in the Ganga–Brahmaputra Delta consist more of sandy and silty matter than clay. This naturally prevents auto-compaction. Hence, the primary cause of the subsidence in the central and coastal Ganga–Brahmaputra Delta can be said to be tectonicity. But the Holocene sediments in the Hooghly estuary contain large amounts of silts and clays, which indicates the possibility of auto-compaction related subsidence. The phenomenon of subsidence is not evident in other parts of the World due which the relative SLR is significantly high in this lower Gangetic delta complex.

SLR has multiple adverse impacts on the lives and livelihoods of Sundarban people. The embankments get damaged (Fig. 3.33) resulting in the intrusion of seawater in the villages adjacent to the estuaries.

Our point is the confusion/noise created on the issue of SLR by factors like 'global warming and subsequent expansion of water,' 'melting of Himalayan glaciers

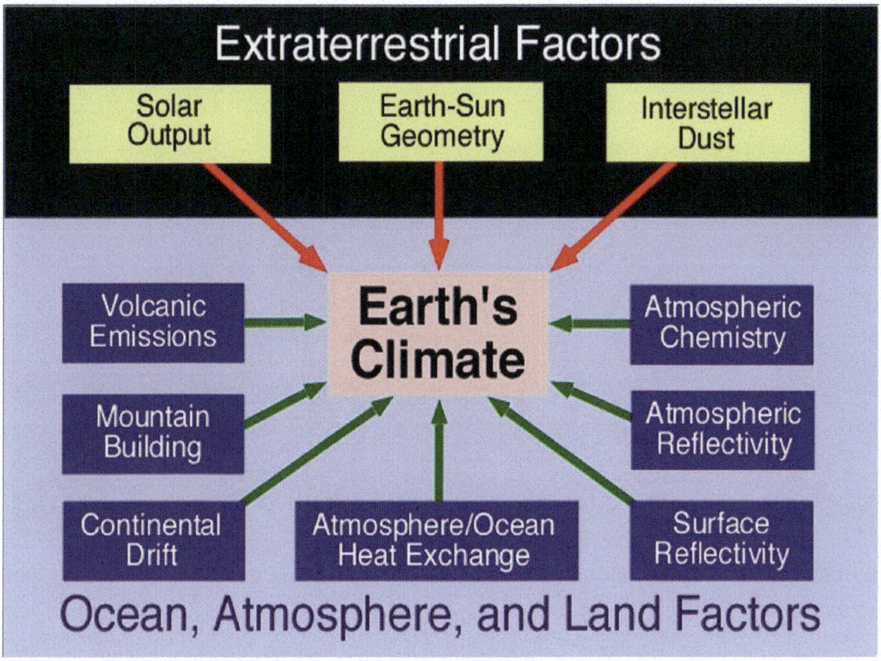

Fig. 3.32 Natural factors influencing the climatic variations in the planet Earth

supplying water to the Ganga-Hooghly estuary' and 'subsidence of land mass'. All these factors result in the rising of sea level and most of them may act in a synergistic way and hence it becomes difficult to pinpoint the exact cause.

3.3.2 Warming of Estuarine Water

Warming of marine and estuarine waters is witnessed across the globe (Table 3.14), but pinpointing the exact cause creates uncertainty. The warming may be due to release of effluents from large number of industries located in the coastal zone, as seen in the case of Hooghly estuary, where a chain of industries have been established from upstream to downstream region or it may be the impact of global warming induced by natural forces.

3.3.3 Alteration of Salinity

In central Indian Sundarbans, we believe that the phenomenon of salinification of Matla estuary clearly indicates SLR as the 'noise' arising of any other external inputs

Fig. 3.33 Damage of embankment at Bali island as observed during the visit of 5th August, 2021

Table 3.14 Increase of surface water temperature in different stations around the world

Location	Rate of increase of surface water temperature	References
The northwest and south west coast of Indian seas	0.2 °C from 1960 to 2005	Vivekanandan et al. (2009)
The southeast coast of Indian seas	0.3 °C from 1960 to 2005	Vivekanandan et al. (2009)
The northern part of the Baltic sea region	0.11 °C/decade from 1871 to 2011	Brohan et al. (2006)
The southern part of the Baltic sea	0.08 °C/decade from 1871 to 2011	Brohan et al. (2006)
The north Atlantic in the latitude belt 40–50° N	0.65 °C between 1920 and 1987	Kushnir (1994)
The Gulf of Bothnia	0.5–0.6 °C/decade from 1970 to 2008	www.hvonstorch.de/klima/pdf/RADOST-BACC.pdf

is practically zero. The input of freshwater in this sector has been blocked since fifteenth century and hence the alteration of salinity is attributed to ingression of seawater from Bay of Bengal in the south. The phenomenon of alteration of salinity due to climate change has been addressed by many researchers, but interferences by other factors often fail to point out the real actor behind climate change scenario. To understand the situation, we can put forward the case study of salinification in central Indian Sundarbans. The increased trend of salinity in the Matla estuary may be due to SLR that brings saline water from Bay of Bengal or it may also be the

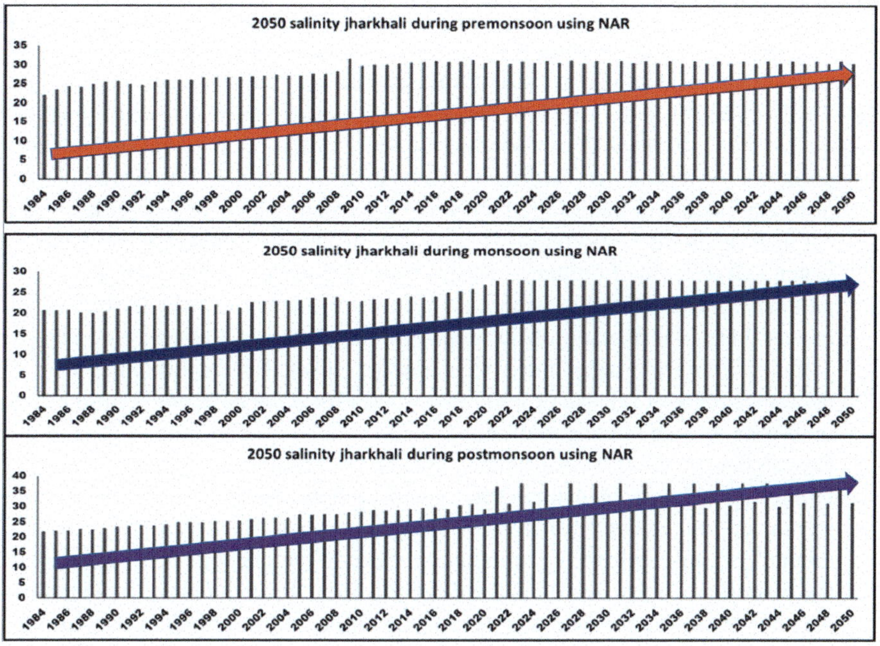

Fig. 3.34 Rising trend of salinity at Jharkhali in central Indian Sundarbans through seasons

effect of heavy siltation in the upstream of this region since the late fifteenth century that has absolutely cut off the input of fresh water in the estuary. The Bidyadhari River which used to drain freshwater to Matla and is now silted and dead (Choudhury and Chaudhury 1994). We monitored the seasonal variation of salinity at Jharkhali (22° 05′ 52.82″ N and 88° 41′ 47.25″ E), a station in central Indian Sundarbans from 1984 to 2022 and observed a rising trend in all the three seasons. The forecasting data also exhibited similar trend till 2050 (Fig. 3.34).

3.4 Take Home Messages

[A] The natural resources present in the islands, adjoining mudflats, estuaries, creeks etc. of Indian Sundarbans are the foundation stones of regional livelihoods. These natural assets encompass both floral and faunal resources, which are exposed to several natural threats like temperature rise, sea level rise, natural disaster, erosion, salinization, acidification, wave actions etc. These natural disasters mostly damage the western and central sectors of Indian Sundarbans due to mass deforestation of mangroves for developing infrastructures related to fish landing, tourism, jetties, shrimp farms, brick kilns etc., which otherwise could act as a bio-shield to protect the super cyclones. The eastern sector, on the other hand, is a Reserve Forest (RF) mostly within the core area where the presence of luxuriant mangrove vegetation not only act as a guard wall against super cyclones, but also arrest the soil particles and prevent erosion.

[B] The livelihood sector in Indian Sundarbans is exposed to several anthropogenic threats like (i) elevated carbon dioxide level in the atmosphere, (ii) overexploitation of fishes, (iii) acidification of estuarine water, (iv) bioaccumulation of heavy metals in the edible fishes, (v) bioaccumulation of pesticides in the edible food products (preferably fishes), (vi) oil and grease level in the estuarine water, (vii) release of untreated wastes from shrimp farms, and (viii) release of untreated wastes from tourism units. Many of these threats have 'noise' which means there is a mixture of anthropogenic and natural sources, e.g., acidification of estuarine water may occur due to elevated carbon dioxide level in the atmosphere or release of untreated wastes from the adjacent shrimp farms, industries, or fish landing stations.

[C] Researches in relation to human impacts on climate change and the interrelationships between natural forces and climate change date back to eighteenth century. However, in most of these researches it is difficult to get a clear 'signal' or cause behind a phenomenon. Several factors appear on the plate of the readers as 'noise' and this creates confusion. To understand the situation, we have presented the case study of salinification in central Indian Sundarbans. The increased trend of salinity in the Matla estuary may be due to SLR that brings saline water from Bay of Bengal or it may also be the effect of heavy siltation in the upstream of this region since the late fifteenth century that has absolutely cut off the input of fresh water in the estuary.

Annexure 3.1: Feedback Questionnaire on Natural Threats in Indian Sundarbans

1. What is your name?

----

2. In which part of Sundarban do you stay? (Tick that which is applicable)

Western ☐
Central ☐
Eastern ☐

3. What is your profession/occupation? (Tick that which is applicable)

   Policy maker ☐
   Researcher ☐
   Fisherman ☐
   Agriculturist ☐
   Others ☐

4. If others, mention it.

   ----

5. What are the major natural threats you witness in your region?

   ----

6. How the natural threats affect your livelihood? (Tick that which is applicable)

   Maximum ☐
   Medium ☐
   Minimum ☐
   No idea ☐

7. How natural disasters (like cyclone) impact your livelihood? (Give score from 1 to 8, where 1 indicates lowest impact and 8 indicates highest impact)

	1	2	3	4	5	6	7	8	
Lowest	○	○	○	○	○	○	○	○	Highest

8. How wave action and tidal surges impact your livelihood? (Give score from 1 to 8, where 1 indicates lowest impact and 8 indicates highest impact)

	1	2	3	4	5	6	7	8	
Lowest	○	○	○	○	○	○	○	○	Highest

9. How temperature rise impacts your livelihood? (Give score from 1 to 8, where 1 indicates lowest impact and 8 indicates highest impact)

	1	2	3	4	5	6	7	8	
Lowest	○	○	○	○	○	○	○	○	Highest

10. How sea level rise impacts your livelihood? (Give score from 1 to 8, where 1 indicates lowest impact and 8 the highest impact)

	1	2	3	4	5	6	7	8	
Lowest	○	○	○	○	○	○	○	○	Highest

11. How salinification impacts your livelihood? (Give score from 1 to 8, where 1 indicates lowest impact and 8 the highest impact)

	1	2	3	4	5	6	7	8	
Lowest	○	○	○	○	○	○	○	○	Highest

12. How acidification impacts your livelihood? (Give score from 1 to 8, where 1 indicates lowest impact and 8 indicates highest impact)

	1	2	3	4	5	6	7	8	
Lowest	○	○	○	○	○	○	○	○	Highest

13. How erosion impacts your livelihood? (Give score from 1 to 8, where 1 indicates lowest impact and 8 indicates highest impact)

	1	2	3	4	5	6	7	8	
Lowest	○	○	○	○	○	○	○	○	Highest

14. How siltation impacts your livelihood? (Give score from 1 to 8, where 1 indicates lowest impact and 8 indicates highest impact)

	1	2	3	4	5	6	7	8	
Lowest	○	○	○	○	○	○	○	○	Highest

15. Do you have any suggestion (s) to combat the natural threat (s) in your region?

--.

References

Agarwal S, Fazli P, Zaman S, Pramanick P, Mitra A (2019) Seasonal variability of acidification in major estuaries of Indian Sundarbans. Glob J Eng Sci Res 6(4):493–498

Araneda M, Perezand PE, Gasca-Leyva E (2008) White shrimp culture is fresh water at three densities: condition state based on length and weight. Aquacult 283:13–18. https://doi.org/10.1016/j.aquaculture.2008.06.030

Botes L (2003) Phytoplankton identification catalogue—Saldanha Bay, South Africa. April 2001, Globallast Monograph Series No.7, IMO London

Brohan P, Kennedy JJ, Harris H, Tett SFB, Jones PD (2006) Uncertainty estimates in regional and global observed temperature changes: a new dataset from 1850. Jr Biophys Res 111:102–106

Chakraborty S, Zaman S, Pramanick P, Raha AK, Mukhopadhyay N, Chakravartty D, Mitra A (2013) Acidification of Sundarbans mangrove estuarine system. Discov Nat 6(14):14–20

Chaudhuri AB, Choudhury A (1994) Mangroves of the Sundarbans. Volume 1: India. World Conservation Union, Gland, p 247

Ferdaushy MH, Alam MM (2015) Length-length and length-weight relationships and condition factor of nine fresh water fish species of Nagesh wari, Bangladesh. Int J Aquacul Biol 3(3):149–154

Froese R (2006) Cube law, condition factor and weight–length relationships: history, meta-analysis and recommendations. J Appl Ichthyol 22(4):241–253. https://doi.org/10.1111/j.1439-0426.2006.00805.x

Fulton TW (1904) The rate of growth of fishes. Twenty-second Annual Report, Part III. Fisheries Board of Scotland, Edinburgh, 141–241

Hazra S, Ghosh T, Das Gupta R, Sen G (2002) Sea level and associated changes in the Sundarbans. Sci C 68:309–321

Hossain MS, Chowdhury SR, Sharifuzzaman SM, Sarker S (2015) Vulnerability of the Bay of Bengal to Ocean acidification. IUCN, International Union for Conservation of Nature, Bangladesh Country Office, Dhaka, Bangladesh, vi+55

Hossain MY, Ahmed ZF, Leunda PM, Jasmine S, Oscoz J, Miranda R (2006) Condition factor, length-weight and length-length relationship of the Asian striped catfish *Mystus vittatus* (Bloch, 1794) (Siluriformes: Bagridae) in the Mathbhanga River, southwestern Bangladesh. J App Ichthyol 22:304–307. https://doi.org/10.1111/j.1439-0426.2006.00801.x

Jana HK, Zaman S, Pramanick P, Mukhopadhyay N, Bose R, Mitra A, Ray Chaudhuri T, Raha AK (2014) Signal of climate change through decadal variation of aquatic pH in Indian Sundarbans. J Mar Sci Res Develop S 11:003. https://doi.org/10.4172/2155-9910.S11-003

Koutrakis ET, Tsikliras AC (2003) Length–weight relationships of fishes from three northern Aegean estuarine systems (Greece). J Appl Ichthyol 19(4):258–260. https://doi.org/10.1046/j.1439-0426.2003.00456.x

Kushnir Y (1994) Interdecadal variations in the North Atlantic Sea surface temperature and associated conditions. J Clim 7:141–157

Mitra A (2013) Sensitivity of mangrove ecosystem to changing climate. Springer, New Delhi, Heidelberg, New York, Dordrecht, London. ISBN-10: 8132215087; ISBN-13:978-8132215080. ISBN 978-81-322-1509-7 (eBook), vol XIX, pp 323. https://doi.org/10.1007/978-81-322-1509-7

Mitra A (2023) Impact of COVID-19 lockdown on environmental health: exploring the situation of the lower Gangetic Delta. Springer eBook ISBN 978-3-031-27242-4, vol XIII, pp 310. https://doi.org/10.1007/978-3-031-27242-4

Mitra A, Gangopadhyay A, Dube A, Andre CKS, Banerjee K (2009) Observed changes in water mass properties in the Indian Sundarbans (Northwestern Bay of Bengal) during 1980–2007. Curr Sci 97(10):1445–1452

Mitra A, Zaman S (2021) estuarine acidification: exploring the situation of mangrove dominated Indian Sundarban Estuaries. Springer eBook ISBN 978-030-84792-0, vol XII, pp 402. https://doi.org/10.1007/978-3-030-84792-0

Mitra A, Zaman S, Pramanick P (2022) Blue economy in Indian Sundarbans: exploring livelihood opportunities. Springer ISBN 978-3-031-07908-5 (e-Book), vol XIV, pp 403. https://doi.org/10.1007/978-3-031-07908-5

Ray Chaudhuri T, Fazli P, Zaman S, Pramanick P, Bose R, Mitra A (2014) Impact of acidification on heavy metals in Hooghly Estuary. J Harmonized Res Appl Sci 2(2):91–97

Roychowdhury R, Vyas P, Zaman S, Roy A, Mitra A (2019) Surface water pH: a proxy to acidification of estuarine water of Indian Sundarbans. Int J Res Anal Rev 6(1):1530–1535

Shabecoff P (1988) Global warming has begun, expert tells senate. New York Times 24 Jun, pA1

Verlencar X (2004) Phytoplankton identification manual. National Institute of Oceanography

Vivekanandan E, Rajagopalan M, Pillai NGK (2009) Recent trends in sea surface temperature and its impact on oil sardine. In: Aggarwal PK (Ed) Impact, adaptation and vulnerability of Indian agriculture to climate change. Ind Coun Agri Res, New Delhi

www.hvonstorch.de/klima/pdf/RADOST-BACC.pdf

Chapter 4
Mangrove-Centric Alternative Livelihoods

Contents

The concept of climate resilient sustainable livelihoods implies not just what people do in order to live and survive comfortably for a short period of time, but the resources that provide them a long-term benefit for several generations even after being exposed to different climatic conditions. In addition to risk factor, different categories of climate resilient sustainable livelihoods take in to account the management of resources coupled with alignment/non-alignment with the existing institutional support and policies related to livelihoods.

In the domain of sustainable livelihood, resources are referred to as 'assets' or 'capitals' and are classified into several sub-categories like human capital (skills, education, and health), physical capital (produced investment goods), financial capital (money savings and access to loan), natural capital (land, water, flora, fauna, microbes etc.) and social capital (networks and association e.g., linkage of the entrepreneurs with the block and panchayet offices in case of Sundarbans). In the sustainable livelihood approach, thrust is given on the asset status of poor individuals or households as fundamental and understanding of the options open to them. The climate resilient sustainable livelihood approach is basically an integrated view of how people survive with evolving social, institutional, political, economic, and environmental/climatic issues. It has proved to have considerable potential in recognizing few important components like (i) the multiple and diverse character of livelihoods (Ellis 1998, 2000), (ii) the prevalence of institutionalized blockages to improving livelihoods, (iii) the social as well as economic character of livelihood strategies, (iv) the principal factors implicated in rising or diminishing vulnerability and (v) the micro-macro (or macro- micro) links that connect varieties of livelihoods to policies.

A. Mitra et al., *Climate Resilient Innovative Livelihoods in Indian Sundarban Delta*,
https://doi.org/10.1007/978-3-031-42633-9_4

In order to develop a sustainable rural livelihood based on endemic natural resources in any type of ecosystem, a sound business model is very essential. This will address the rationale of how an organization creates, delivers, and captures value. It encompasses several routes of value creation (i.e., economic, social, cultural, and ecological). As such a sustainable model attempts to link business activity to spatial level like developing more balanced and sustainable rural-urban relations.

Again, to develop a climate resilient sustainable business model in the frame work of mangrove dominated Indian Sundarbans, a wide range of relevant ecological data sets need to be explored considering the physico-chemical features of three major sectors (western, central, and eastern) of Indian Sundarbans as pointed here.

(I) Regional, sectoral, rural, and urban employment dynamics.
(II) Sector-wise ecological profile of Indian Sundarbans (preferably salinity-based spatial variation).
(III) Sector-wise existing livelihoods in Indian Sundarbans.
(IV) Awareness on climate resilient mangrove-centric innovative livelihood.
(V) Sector-wise island business model prospects, and
(VI) Existing institutional mechanism and policies related to livelihoods in the islands of Indian Sundarbans.

Based on contrasting spatial variations in ecological characteristics of Indian Sundarbans, we have discussed in this chapter few innovative livelihoods primarily based on the salinity of the region.

4.1 Alternative Livelihoods for High Saline Zone

The population of the planet Earth is increasing at an alarming rate. A total of 8,03,71,21,208 individuals has been recorded in recent times (Table 4.1), and the figure is expected to touch 9.9 billion in 2050. Interestingly it is observed that India has exceeded China in recent times in terms of population. In this context, it is the need of the hour to provide economic, food, health, and environmental securities to this massive mass of population.

Today, approximately 3 billion people, which is almost near to half of the World's population live within 200 km of the World's coastline. The bitter truth of the present *era* referred to as the Sea Level Rise (SLR) has posed severe adverse impacts on this large coastal population by way of accelerating coastal erosion, inundations, storm floods, tidal waters encroachment into estuaries and river systems, contamination of freshwater reserves and food crops/agricultural land (with saline water), loss of properties etc. Latest figures exhibit that 90% of the disasters are linked with climate change that incurs a loss of world economy by US$520 billion per year and has high probability to push about 26 million people into poverty or below the poverty line

Table 4.1 Country-wise World population (on 12th June, 2023 at 12.35 am GMT)

Sl. No.	Country	Population
1	India	1,42,79,41,710
2	China	1,42,56,89,878
3	United States	33,99,01,310
4	Indonesia	27,74,14,933
5	Pakistan	24,02,40,558
6	Nigeria	22,35,29,201
7	Brazil	21,63,58,015
8	Bangladesh	17,28,63,001
9	Russia	14,44,70,263
10	Mexico	12,84,06,547
11	Ethiopia	12,63,61,802
12	Japan	12,33,28,778
13	Philippines	11,72,45,394
14	Egypt	11,26,26,263
15	Dr Congo	10,20,90,664
16	Vietnam	9,88,24,849
17	Iran	8,91,38,666
18	Turkey	8,57,92,755
19	Germany	8,32,96,600
20	Thailand	7,17,96,361
21	United Kingdom	6,77,24,998
22	Tanzania	6,73,36,131
23	France	6,47,50,026
24	South Africa	6,03,86,132
25	Italy	5,88,79,615
26	Kenya	5,50,43,860
27	Myanmar	5,45,57,504
28	Colombia	5,20,72,052
29	South Korea	5,17,86,026
30	Uganda	4,85,12,985
31	Sudan	4,80,44,411
32	Spain	4,75,21,923
33	Argentina	4,57,58,965
34	Algeria	4,55,70,740
35	Iraq	4,54,51,769
36	Afghanistan	4,21,81,325
37	Poland	4,10,72,956
38	Canada	3,87,64,241

(continued)

Table 4.1 (continued)

Sl. No.	Country	Population
39	Morocco	3,78,20,370
40	Saudi Arabia	3,69,19,318
41	Ukraine	3,66,74,465
42	Angola	3,66,26,493
43	Uzbekistan	3,51,36,729
44	Yemen	3,44,10,478
45	Peru	3,43,35,505
46	Malaysia	3,42,89,507
47	Ghana	3,40,88,212
48	Mozambique	3,38,47,842
49	Nepal	3,08,78,556
50	Madagascar	3,02,88,024
51	Ivory Coast	2,88,35,326
52	Venezuela	2,88,09,153
53	Cameroon	2,86,08,766
54	Niger	2,71,50,216
55	Australia	2,64,25,503
56	North Korea	2,61,56,230
57	Taiwan	2,39,21,800
58	Mali	2,32,56,810
59	Burkina Faso	2,32,21,155
60	Syria	2,31,68,157
61	Sri Lanka	2,18,90,628
62	Malawi	2,09,04,044
63	Zambia	2,05,40,554
64	Romania	1,99,09,699
65	Chile	1,96,28,114
66	Kazakhstan	1,95,95,649
67	Chad	1,82,49,385
68	Ecuador	1,81,80,647
69	Somalia	1,81,14,523
70	Guatemala	1,80,78,910
71	Senegal	1,77,39,555
72	Netherlands	1,76,15,512
73	Cambodia	1,69,35,481
74	Zimbabwe	1,66,47,047
75	Guinea	1,41,73,234
76	Rwanda	1,40,78,124

(continued)

Table 4.1 (continued)

Sl. No.	Country	Population
77	Benin	1,36,93,974
78	Burundi	1,32,20,361
79	Tunisia	1,24,52,485
80	Bolivia	1,23,79,226
81	Haiti	1,17,17,385
82	Belgium	1,16,84,664
83	Jordan	1,13,34,757
84	Dominican Republic	1,13,27,726
85	Cuba	1,11,95,433
86	South Sudan	1,10,79,287
87	Sweden	1,06,08,807
88	Honduras	1,05,85,109
89	Czech Republic	1,04,94,803
90	Azerbaijan	1,04,09,864
91	Greece	1,03,43,244
92	Papua New Guinea	1,03,20,258
93	Portugal	1,02,48,917
94	Hungary	1,01,65,584
95	Tajikistan	1,01,33,706
96	United Arab Emirates	95,12,936
97	Belarus	95,00,533
98	Israel	91,67,470
99	Togo	90,43,142
100	Austria	89,57,976
101	Switzerland	87,93,718
102	Sierra Leone	87,81,419
103	Laos	76,28,369
104	Hong Kong	74,91,281
105	Serbia	71,52,356
106	Nicaragua	70,41,228
107	Libya	68,84,453
108	Paraguay	68,57,097
109	Kyrgyzstan	67,29,937
110	Bulgaria	66,92,144
111	Turkmenistan	65,11,837
112	El Salvador	63,63,304
113	Republic Of The Congo	60,99,819
114	Singapore	60,12,756

(continued)

Table 4.1 (continued)

Sl. No.	Country	Population
115	Denmark	59,09,437
116	Slovakia	58,00,609
117	Central African Republic	57,33,626
118	Finland	55,45,311
119	Norway	54,72,229
120	Liberia	54,12,311
121	Palestine	53,64,836
122	Lebanon	53,61,308
123	New Zealand	52,25,969
124	Costa Rica	52,10,370
125	Ireland	50,55,132
126	Mauritania	48,56,267
127	Oman	46,40,777
128	Panama	44,64,972
129	Kuwait	43,07,977
130	Croatia	40,09,765
131	Eritrea	37,45,458
132	Georgia	37,29,102
133	Mongolia	34,44,698
134	Moldova	34,42,325
135	Uruguay	34,23,108
136	Puerto Rico	32,59,822
137	Bosnia And Herzegovina	32,11,831
138	Albania	28,32,931
139	Jamaica	28,25,544
140	Armenia	27,77,970
141	Gambia	27,69,561
142	Lithuania	27,19,664
143	Qatar	27,15,243
144	Botswana	26,73,057
145	Namibia	26,02,205
146	Gabon	24,34,107
147	Lesotho	23,29,006
148	Guinea Bissau	21,48,547
149	Slovenia	21,19,675
150	North Macedonia	20,86,007
151	Latvia	18,31,195
152	Equatorial Guinea	17,12,540

(continued)

Table 4.1 (continued)

Sl. No.	Country	Population
153	Trinidad And Tobago	15,34,773
154	Bahrain	14,84,853
155	Timor Leste	13,59,612
156	Estonia	13,22,929
157	Mauritius	13,00,557
158	Cyprus	12,59,646
159	Eswatini	12,10,330
160	Djibouti	11,35,635
161	Reunion	9,81,468
162	Fiji	9,36,047
163	Comoros	8,51,255
164	Guyana	8,13,506
165	Bhutan	7,87,096
166	Solomon Islands	7,39,604
167	Luxembourg	6,54,440
168	Montenegro	6,26,485
169	Suriname	6,22,908
170	Cape Verde	5,98,354
171	Western Sahara	5,86,603
172	Malta	5,34,900
173	Maldives	5,21,185
174	Brunei	4,52,360
175	Bahamas	4,12,459
176	Belize	4,10,497
177	Guadeloupe	3,95,839
178	Iceland	3,75,154
179	Martinique	3,66,981
180	Mayotte	3,35,503
181	Vanuatu	3,34,178
182	French Guiana	3,11,827
183	Barbados	2,81,995
184	Sao Tome And Principe	2,31,692
185	Samoa	2,25,517
186	Saint Lucia	1,80,251
187	Kiribati	1,33,351
188	Grenada	1,26,183
189	Micronesia	1,15,224
190	Tonga	1,07,773

(continued)

Table 4.1 (continued)

Sl. No.	Country	Population
191	Seychelles	1,07,660
192	Saint Vincent And The Grenadines	1,03,698
193	Antigua And Barbuda	94,298
194	Andorra	80,088
195	Dominica	73,040
196	Greenland	56,643
197	Saint Kitts And Nevis	47,755
198	Marshall Islands	41,996
199	Liechtenstein	39,584
200	Monaco	36,297
201	San Marino	33,642
202	Palau	18,058
203	Nauru	12,780
204	Tuvalu	11,396
205	Vatican City	518

Source https://worldpopulationreview.com/countries

every year (https://www.un.org/sustainabledevelopment/wp-content/uploads/2017/05/Ocean-fact-sheet-package.pdf). The delta complex of Indian Sundarbans at the apex of Bay of Bengal is no exception to these dark facts. Several islands have shown massive erosion after the event of super—cyclone Aila that hit the lower Gangetic delta encompassing Indian Sundarbans between 25 to 27th May, 2009 (Figs. 4.1, 4.2, 4.3, 4.4, 4.5, 4.6, 4.7, 4.8 and 4.9).

The central sector of Indian Sundarbans is known for hypersalinity because of Bidyadhari siltation since fifteenth century (Chaudhuri and Choudhury 1994; Mitra 2013). We have monitored the aquatic salinity profile in few islands in this sector where we observed an increasing trend since the last three decades (Table 4.2).

Based on this background few innovative high saline-based livelihoods befitted for Indian Sundarbans are discussed here in brief.

4.1.1 **Suaeda** *and* **Salicornia** *Farming in Supra-Littoral Zone*

The rate of food production in the planet Earth needs to be increased by around 70% to keep pace with the predicted population of 2050. In coastal areas, this is a great challenge as the ongoing rate of sea level rise causes salinization of both soil and water and in many cases the agricultural land of the island villages. According to Flowers et al. (1997), about 7% of the Earth's soil has become salty in nature. Under this situation, farming of halophytes as sources of food and vegetables can be a

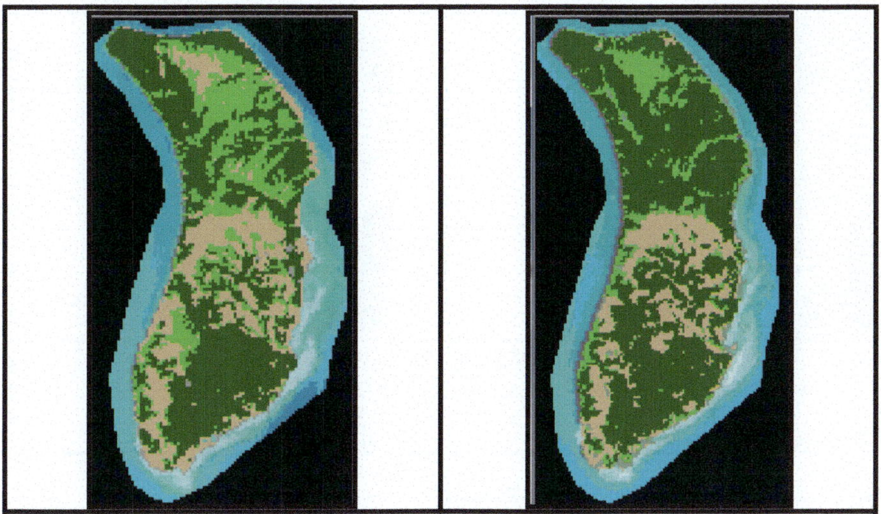

MANGROVE VEGETATION = 2527 ha	MANGROVE VEGETATION = 2614 ha
TOTAL ISLAND AREA = 3356 ha	TOTAL ISLAND AREA = 3247 ha
PRE-AILA PHASE	POST AILA PHASE
PERIOD OF IMAGERY: DEC 2008	PERIOD OF IMAGERY: JAN 2010

Fig. 4.1 Lothian Island. *Source* Mitra and Zaman (2015)

MANGROVE VEGETATION = 6270 ha	MANGROVE VEGETATION = 6359 ha
TOTAL ISLAND AREA = 6899 ha	TOTAL ISLAND AREA = 6931 ha
PRE-AILA PHASE	POST AILA PHASE
PERIOD OF IMAGERY: DEC 2008	PERIOD OF IMAGERY: JAN 2010

Fig. 4.2 Baghmara Island. *Source* Mitra and Zaman (2015)

sustainable roadmap to meet the nutritional demand of the rapidly rising population. List of few important halophytes that can be grown in saline soil are presented in Table 4.3.

Among all the species highlighted in Table 4.3, *Suaeda* and *Salicornia* are widely available in Indian Sundarbans.

MANGROVE VEGETATION = 6068 ha	MANGROVE VEGETATION = 5725 ha
TOTAL ISLAND AREA = 6825 ha	TOTAL ISLAND AREA = 6645 ha
PRE-AILA PHASE	POST AILA PHASE
PERIOD OF IMAGERY: DEC 2008	PERIOD OF IMAGERY: JAN 2010

Fig. 4.3 Gona Island. *Source* Mitra and Zaman (2015)

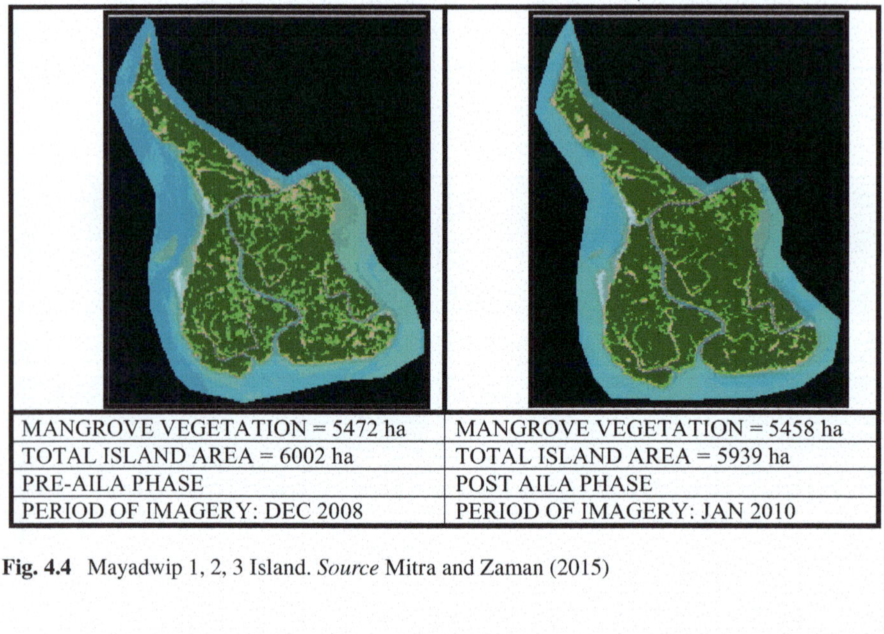

MANGROVE VEGETATION = 5472 ha	MANGROVE VEGETATION = 5458 ha
TOTAL ISLAND AREA = 6002 ha	TOTAL ISLAND AREA = 5939 ha
PRE-AILA PHASE	POST AILA PHASE
PERIOD OF IMAGERY: DEC 2008	PERIOD OF IMAGERY: JAN 2010

Fig. 4.4 Mayadwip 1, 2, 3 Island. *Source* Mitra and Zaman (2015)

Suaeda maritima is a mangrove associate plant species abundantly available in the supra-littoral zone of mangrove forest that grow in the high saline habitat (Fig. 4.10). The tender leaves of the species are used as fresh vegetable and are also consumed in cooked form. The cooked seablite (common name of *Suaeda maritima*) is quite salty, and hence cooked with other types of vegetable to reduce the salty taste (Tanaka 1976). Local people in Samut Songkram province use seablite for different types of cooking such as traditional seablite salad, seablite curry with crabs, or scalded

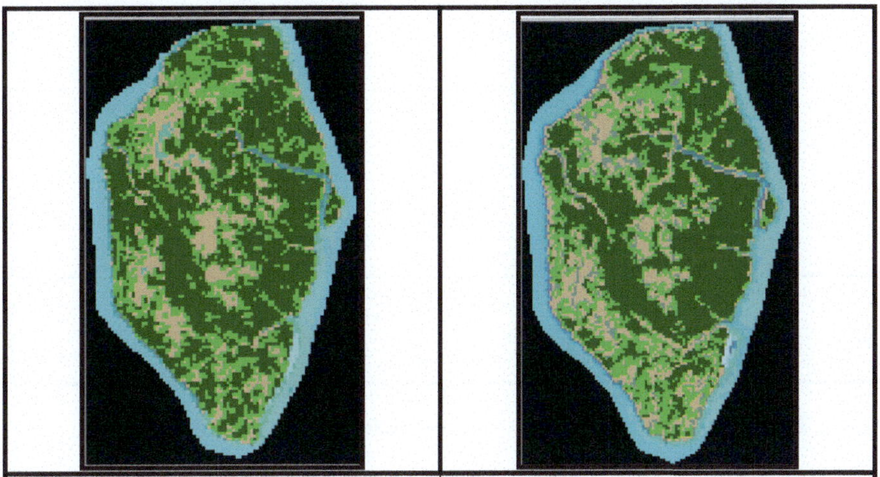

MANGROVE VEGETATION = 2851 ha	MANGROVE VEGETATION = 2544 ha
TOTAL ISLAND AREA = 3332 ha	TOTAL ISLAND AREA = 3164 ha
PRE-AILA PHASE	POST AILA PHASE
PERIOD OF IMAGERY: DEC 2008	PERIOD OF IMAGERY: JAN 2010

Fig. 4.5 Thakuran (Dhanchi) Island. *Source* Mitra and Zaman (2015)

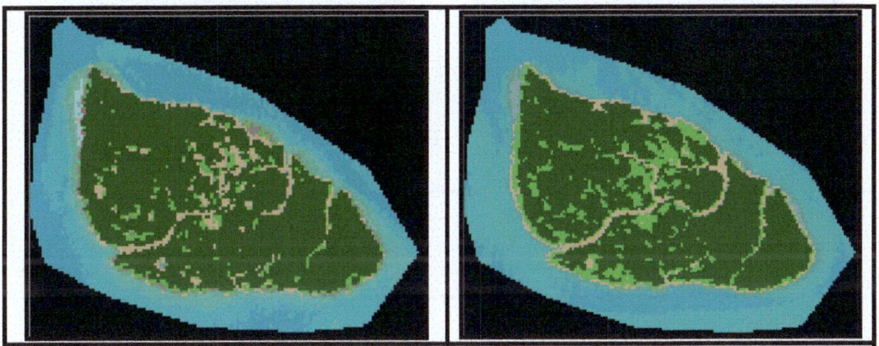

MANGROVE VEGETATION = 2335 ha	MANGROVE VEGETATION = 2333 ha
TOTAL ISLAND AREA = 2599 ha	TOTAL ISLAND AREA = 2550 ha
PRE-AILA PHASE	POST AILA PHASE
PERIOD OF IMAGERY: DEC 2008	PERIOD OF IMAGERY: JAN 2010

Fig. 4.6 Mayadwip 4,5 Island. *Source* Mitra and Zaman (2015)

seablite with chilli paste. The edible part is the young leaves which are scalded for about 10–15 min and then knocked with cold water to make them crispier (Pornpitakdamrong and Sudjaroen 2014). In the South Indian states, seablite is pickled in vinegar or used for cooking as well as domestic animal food (Bandaranayke 2002).

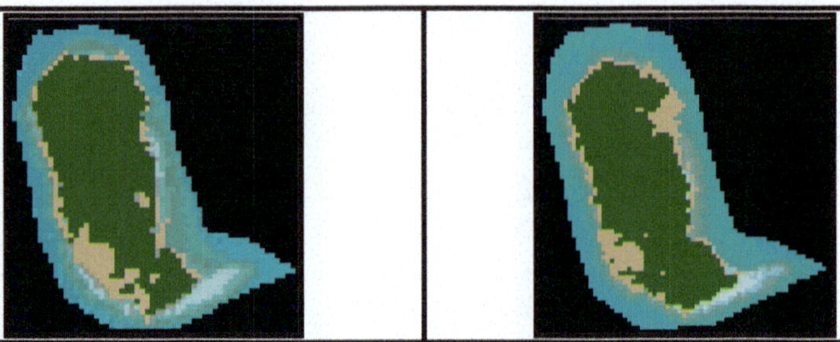

MANGROVE VEGETATION = 368 ha	MANGROVE VEGETATION = 346 ha
TOTAL ISLAND AREA = 449 ha	TOTAL ISLAND AREA = 429 ha
PRE-AILA PHASE	POST AILA PHASE
PERIOD OF IMAGERY: DEC 2008	PERIOD OF IMAGERY: JAN 2010

Fig. 4.7 Jambudwip Island. *Source* Mitra and Zaman (2015)

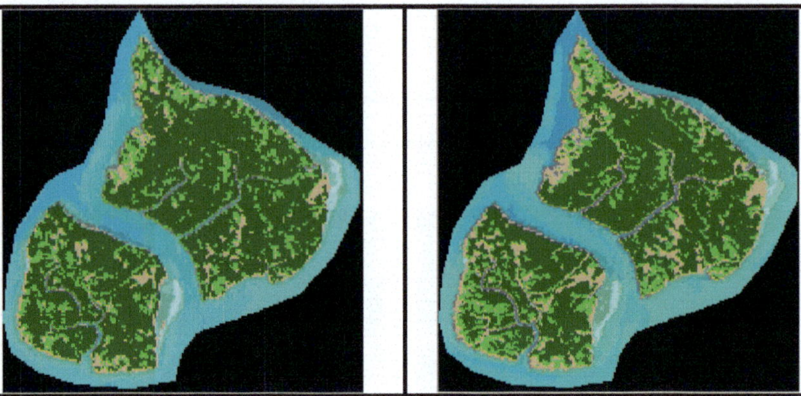

MANGROVE VEGETATION = 5332 ha	MANGROVE VEGETATION = 5197 ha
TOTAL ISLAND AREA = 5894 ha	TOTAL ISLAND AREA = 5845 ha
PRE-AILA PHASE	POST AILA PHASE
PERIOD OF IMAGERY: DEC 2008	PERIOD OF IMAGERY: JAN 2010

Fig. 4.8 Chulkati Island. *Source* Mitra and Zaman (2015)

S. maritima can grow naturally in high saline soil, and so it can be considered as a climate resilient low-cost vegetable with high nutritional value (Figs. 4.11 and 4.12). There should be a promotion of seablite as ready-made edible products for the convenience of the consumers and thus can be a source of climate resilient alternative livelihood for the Sundarban island dwellers.

Salicornia brachiata is also a mangrove associate species, which is commonly known as pickleweed, glasswort, sea beans, sea asparagus, crow's foot greens, and samphire. It is a halophyte, belonging to Amaranthaceae family (Singh et al. 2014) (Fig. 4.13). The term *Salicornia* has originated from the Latin word meaning 'salt'. Several studies report that some species, for example *Salicornia europaea* show

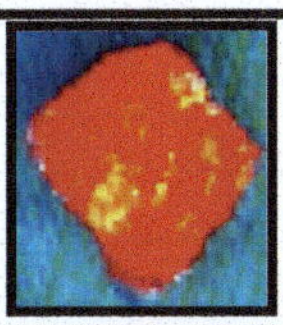

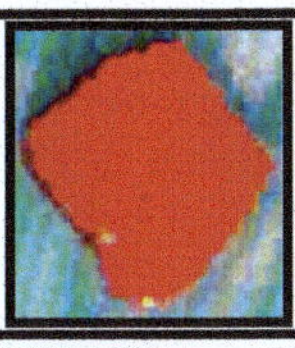

TOTAL ISLAND AREA = 426 ha	TOTAL ISLAND AREA = 436 ha
PRE-AILA PHASE	POST AILA PHASE
PERIOD OF IMAGERY: DEC 2008	PERIOD OF IMAGERY: JAN 2010

Fig. 4.9 Ghoramara Island. *Source* Mitra and Zaman (2015)

tolerance towards salinity as high as 3% NaCl (Yamamoto et al. 2009). This fleshy plant grows at the edges of wetlands, marshes, sea shores, and mudflats, on substratum with high alkalinity (Smillie 2015).

Salicornia has been historically used for both non-edible and edible purposes. Usage of the plant as a source of soda (sodium carbonate) for glass making dates to centuries. Oriental pharmacopeia reports its medicinal uses. The efficacy of *Salicornia herbacea* against oxidative stress, inflammation, diabetes, asthma, hepatitis, cancer, gastroenteritis has been reported (Essaidi et al. 2013). Food use was not altogether new, with multiple reports of their consumption as a salt source. However, recent dearth in food availability, quest for sustainable food sources, climate change induced vulnerability, extinction of food crops due to salinization of agricultural lands and foraging interest has brought this genus on the surface of the climate resilient livelihoods. This plant's aerial parts are consumed in salads or processed into pickles, beverages etc. The present discussion has great relevance in context to the economic profile of Indian Sundarbans as limitation of traditional natural resources (like fishes, honey, fuel, fodder, fire wood etc.) for meeting the daily needs has become a major issue for coastal population and island dwellers. This issue has become very acute in case of Indian Sundarbans where the population is around 4.2 million. Majority of this population illegally intrude into the mangrove forest to procure fish, honey, wood, fire wood, fodder etc., which not only put them in risk (because of man-animal conflict), but also damage the natural resource reservoir of the ecosystem. A major part of the population also gets engaged in the wild harvest of prawn seeds (Fig. 4.14) due to which juveniles of several other commercially important fish species get damaged and destructed (Chowdhury et al. 2017; Mitra et al. 2022).

In addition, food insecurity, natural disasters, diseases, malnutrition etc. are the common associates of Sundarban people. Under this scenario, farming of mangrove associate species like *Suaeda* or *Salicornia* can serve as an effective road map to get rid from these negative issues and open a new horizon in the domain of alternative livelihood. The species has considerable nutritional value.

Halophyte farming (preferably *Suaeda* and *Salicornia*) can provide a wide range of ecosystem services (Fig. 4.15).

Table 4.2 Surface water salinity (in psu) during three seasons around Thakuran Island (21° 49′ 53.17″ N and 88° 31′ 25.57″ E) during 1984–2022

Year	Premonsoon	Monsoon	Postmonsoon
1984	20.05 ± 1.11	18.66 ± 1.10	19.79 ± 1.21
1985	21.56 ± 1.14	18.73 ± 1.10	20.06 ± 1.22
1986	22.32 ± 1.16	18.79 ± 1.11	20.17 ± 1.22
1987	22.19 ± 1.16	18.13 ± 1.07	20.66 ± 1.23
1988	23.02 ± 1.17	18.02 ± 1.02	20.38 ± 1.23
1989	23.66 ± 1.17	18.33 ± 1.11	21.06 ± 1.38
1990	23.83 ± 1.17	19.01 ± 1.16	21.51 ± 1.42
1991	22.99 ± 1.16	19.66 ± 1.20	21.55 ± 1.44
1992	22.69 ± 1.16	19.73 ± 1.20	21.94 ± 1.46
1993	23.56 ± 1.17	19.85 ± 1.22	21.74 ± 1.47
1994	23.98 ± 1.17	19.73 ± 1.20	22.19 ± 1.48
1995	24.02 ± 1.17	19.98 ± 1.22	22.97 ± 1.49
1996	24.05 ± 1.17	19.59 ± 1.20	23.04 ± 1.49
1997	24.83 ± 1.18	20.0 ± 1.24	22.9 ± 1.49
1998	24.79 ± 1.18	20.05 ± 1.24	23.41 ± 1.49
1999	24.73 ± 1.18	18.63 ± 1.26	23.39 ± 1.49
2000	24.95 ± 1.18	19.35 ± 1.31	23.49 ± 1.49
2001	25.02 ± 1.18	20.59 ± 1.38	23.94 ± 1.49
2002	25.11 ± 1.20	20.91 ± 1.38	24.40 ± 1.50
2003	25.43 ± 1.20	20.97 ± 1.40	24.47 ± 1.51
2004	25.18 ± 1.20	21.02 ± 1.40	24.43 ± 1.51
2005	25.16 ± 1.20	21.15 ± 1.41	25.54 ± 1.60
2006	25.84 ± 1.25	21.77 ± 1.43	25.36 ± 1.60
2007	25.66 ± 1.25	21.83 ± 1.43	25.67 ± 1.60
2008	26.32 ± 1.22	21.91 ± 1.43	25.70 ± 1.60
2009	29.66 ± 1.67	20.86 ± 1.42	26.29 ± 1.62
2010	27.85 ± 1.55	20.95 ± 1.43	26.52 ± 1.62
2011	28.11 ± 1.60	21.37 ± 1.44	26.87 ± 1.62
2012	28.05 ± 1.60	21.51 ± 1.44	26.70 ± 1.62
2013	28.44 ± 1.62	21.68 ± 1.45	26.90 ± 1.62
2014	28.71 ± 1.68	22.02 ± 1.46	27.27 ± 1.66
2015	28.94 ± 1.68	21.88 ± 1.45	27.78 ± 1.66
2016	29.06 ± 1.71	22.02 ± 1.46	27.86 ± 1.67
2017	28.34 ± 1.45	21.07 ± 1.28	26.44 ± 1.39
2018	29.20 ± 1.68	23.06 ± 1.33	27.98 ± 1.44
2019	29.83 ± 1.51	24.01 ± 1.26	28.01 ± 1.46
2020	29.90 ± 1.71	25.01 ± 1.33	28.38 ± 1.29
2021	31.20 ± 1.54	25.56 ± 1.81	28.94 ± 1.21
2022	31.67 ± 1.80	26.06 ± 1.55	29.05 ± 1.62

Table 4.3 Halophytic species used as food and vegetables around the world

Plant	Historical and traditional uses	Plant part	Nutritional factors	Where	References
Eelgrass (*Zostera marina*)	Bread; consumed by Seri Indians	Seeds/grain	Seed contains about 50% starch, 13% protein, and 1% fat	South-western United States	Felger and McRoy (1975, 1976)
Palmer salt grass (*Distichlis palmeri*)	Bread; consumed as gruel (a thinner version of porridge) by Yuman and Cocopah Indians	Seeds/grain	Seed contains 79.5% carbohydrate, 7–8% protein, 8.4% fibre and 1.8% fat	Gulf of California, United States	Yensen et al. (1985), Pearlsteina et al. (2012)
Pearl millet (*Pennisetum typhoidea*)	Staple food of native people	Seeds/grain	Gluten free and rich source of energy,[a] also rich in vitamins B	Sub-Saharan Africa and northwest India	BOSTID (1990), Nambiar et al. (2011)
Quinoa (*Chenopodium quinoa*)	Soup, bread, cake, brewing beer, cooked green vegetable by native people	Seeds	Balanced in amino acids[b]	Andean highlands in Peru	Risi and Galway (1984), Koziol (1992), Adolf et al. (2013)
Mesquite (*Prosopis glandulosa*)	Food consumed by Seri Indians	Pods and seeds	Protein content 39.9%	Gulf of California	Felger and Moser (1976)
Saltwort (*Batis maritima*)	Food eaten raw or cooked or pickled by the Seri Indians	Roots and stems	Essential amino acids, tocopherol, antioxidants	South-western United States	Felger and Moser (1976), Debez et al. (2010)
Seaside purslane (*Sesuvium portulacastrum*)	Consumed as a vegetable	Leaves and stems	Edible portion shows high values for calcium, iron, and carotene	India, Indonesia and southern China	BOSTID (1990), Lokhande et al. (2009)
Salicornia and *Sarcocornia* spp.	Salad greens, vegetable	Leaves	Good source of vitamin A, minerals, fatty acids, polyphenol	USA and European fresh markets	Zerai et al. (2010); Ventura et al. (2011)

(continued)

Table 4.3 (continued)

Plant	Historical and traditional uses	Plant part	Nutritional factors	Where	References
Tetragonia tetragonioides	Frozen like spinach	Leaves	Amino acids, antioxidants	First white settlers in Australia	Ahmed and Johnson (2000), Słupski et al. (2010)
Sea Fennel (*Crithmum maritimum*)	Food (salad) by native people; used to protect sailors from scurvy (fresh and pickled)	Leaves	High content of vitamin C	Italy; Greek islands; British Isles	Frank (1982), Ben Hamed et al. (2004)
Mountain spinach (*Atriplex hortensis*)	Leaves for salad	Leaves	High in protein	France and Central Europe	Carlsson and Clarke (1983), Shannon and Grieve (1999),
Common purslane (*Portulaca oleracea*)	Salads and soup	Fresh leaves and stem	Omega-3 fatty acids, vitamin C, vitamin A	Mediterranean region and Central Europe; Pacific Northwest of the U.S	Shannon and Grieve (1999), Simopoulos (2004), Yazici et al. (2007)
Scurvy grass (*Cochlearia officinalis*)	As salad; used by the Dutch mariners to combat the scurvy	Fresh leaf and stem	Vitamin C, glucosinolates	Netherlands	Maat (2004), de Vos et al. (2013)

Note [a]The energy value of pearl millet is 361 kcal/100 g which is comparable with wheat (346 kcal/100 g) and rice (345 kcal/100 g), [b](Particularly histidine and lysine) and their composition is better balanced, than in major cereals

Fig. 4.10 *Suaeda maritima* on the mudflats in Indian Sundarbans

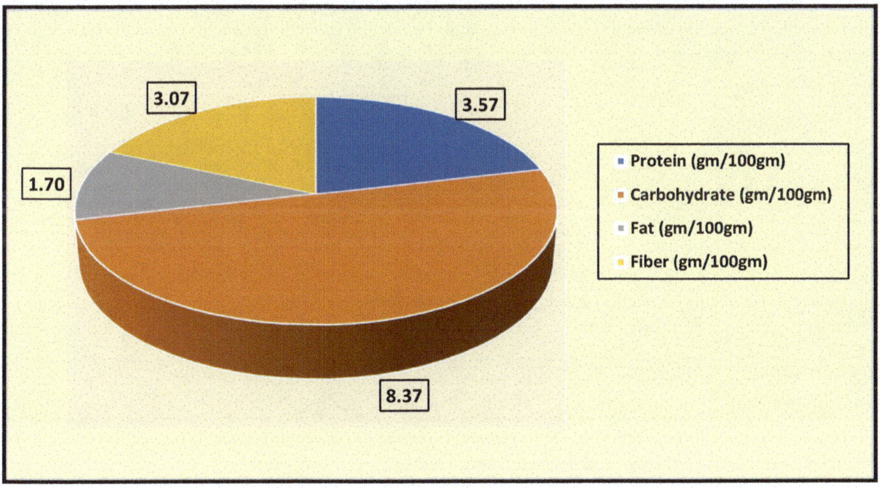

Fig. 4.11 Protein, carbohydrate, fat, and fiber content in *Suaeda maritima*

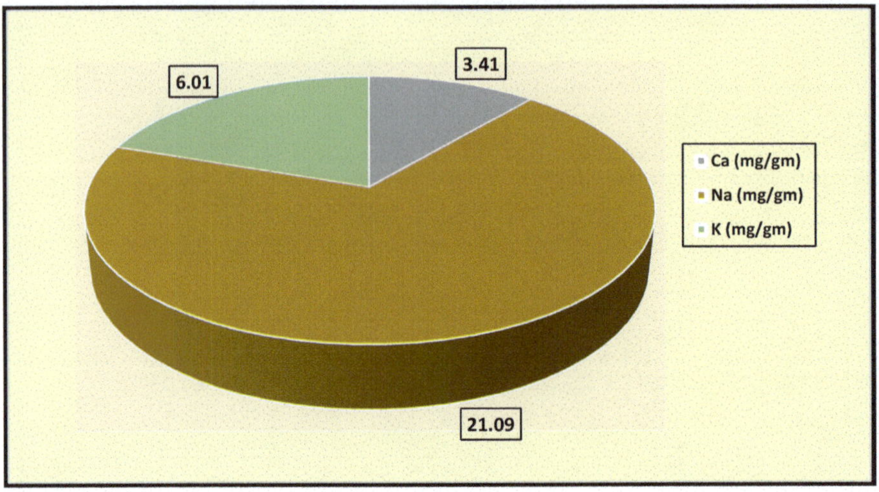

Fig. 4.12 Ca, Na, and K content in *Suaeda maritima*

Fig. 4.13 *Salicornia brachiata* in Indian Sundarbans

We observed that *Suaeda maritima* and *Salicornia brachiata* when cultivated in high saline environment yield good results (in terms of biomass), which indicates that there is scope to grow these halophytes in high saline soil or sea water inundated

Fig. 4.14 Illegal prawn seed collection from the estuarine water during dusk serves as a source of livelihood for the Sundarban population

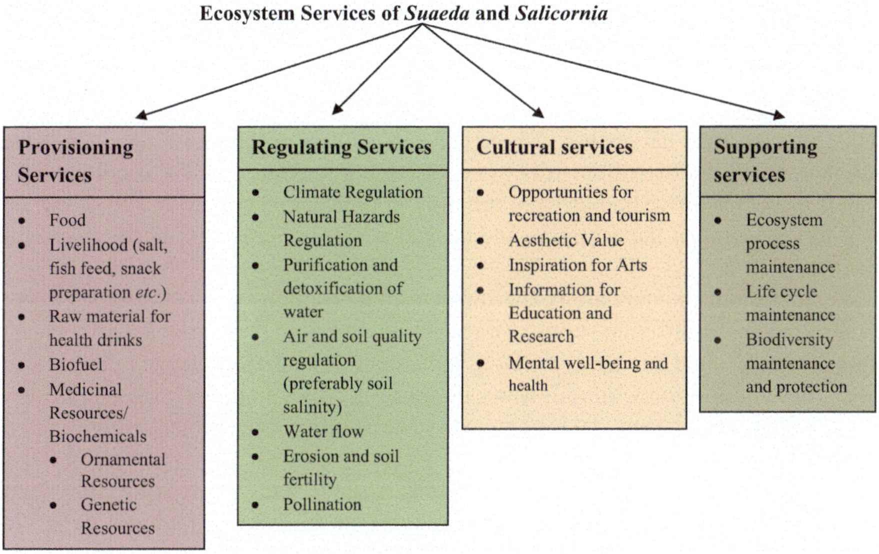

Fig. 4.15 Ecosystem services of *Suaeda* and *Salicornia*

region. The soil usually has high sodium and potassium level. Large scale farming of these mangrove associate species can help retard the pace of salinity rise. The field level technology to grow these mangrove associate species is highlighted here in steps.

1st step

In the 1st step *Suaeda maritima* and *Salicornia brachiata* are raised in the nursery for 45 days in 3″ × 5″ polythene bags filled with garden soil. Shade nets (50%) are commonly used to reduce the light and temperature. Fresh water/water with salinity around 2 psu is used to raise the saplings in the nursery.

2nd step

In this step the 45-day-old seedlings are transplanted and each species are grown in the saline land preferably in the supra-littoral zone. Initially fresh water/water with salinity around 2–4 psu is used for irrigation (2 wettings) and then brackish water from nearby creeks is used.

Remarks

Since these mangrove associate plants are highly saline- and- drought-tolerant, two wettings with brackish water are given. The salinity of the brackish water used for the first wetting is around 25–38 psu, whereas salinity of the brackish water given for the second wetting is around 20–24 psu.

3rd step

In this step the final harvesting is done before the onset of the monsoon since these halophytes will die thereafter. It is to be noted in this context that these mangrove associate species thrive and grow luxuriantly in high saline condition and die when the dilution factor of the estuarine and coastal water increases during the monsoon season.

We carried out an interesting experiment during postmonsoon 2022 on the growth of these species at different salinity (through dilution with proportionate rain water) using Above Ground Biomass (AGB) as indicator (Table 4.4) and observed that with the increase of dilution with rain water the AGB decreased almost proportionately.

A detailed enterprise model in context to sustainability of the project is highlighted in Table 4.5.

Halophyte farming in the framework of Indian Sundarbans involves the cultivation of *Suaeda* and *Salicornia* that can provide job opportunities in multiple tiers. We also carried out a SWOT (Strength, Weakness, Opportunity, and Threat) analysis to evaluate the viability of this climate—resilient alternative livelihood in the framework of Indian Sundarbans (Tables 4.6, 4.7, 4.8 and 4.9). To obtain data for SWOT analysis, stratified random sampling technique was employed and data was collected from 980 respondents comprising of 370 fisherman, 262 agriculturists, 167 hotel/tourism unit owners and 181 researchers using questionnaires.

Table 4.4 Biomass of *Suaeda maritima* and *Salicornia brachiata* at different experimental plots with variable salinity (in psu) during 2022

Species	Total biomass (kgm^{-2})					
	30 psu	25 psu	20 psu	15 psu	10 psu	5 psu
 *Suaeda maritima*	4.98	4.02	3.66	2.12	1.15	Mortality
 *Salicornia brachiata*	3.88	3.12	2.93	2.08	1.97	Mortality

**Note* Variation in salinity was carried out by mixing appropriate volume of rain water through pump run by generator from the rain water harvested pond

In Tables 4.10a, 4.10b, 4.10c and 4.10d, we present an investment profile for farming these halophyte species to achieve a sustainable climate—resilient agribusiness that may have high impact on the economic and environmental profiles of Sundarban delta.

Supply chain has got immense importance for sustaining a business and acts like an invisible conveyor belt that keeps the business moving smoothly and reaches the product(s) from the producer to the end level consumer. The supply chain related to *Suaeda*/*Salicornia* production is highlighted in Fig. 4.16. This chain shows the distribution of the halophytes from the production centre (halophyte farm in the high saline zone) to the consumers involving the traders and wholesalers in the loop. Quality control and packaging are also the important components of this halophytes-based supply chain proposed for the high saline region of Indian Sundarbans.

4.1.2 Salicornia-*Based Shrimp Feed*

Culture of *Penaeus vannamei* is undoubtedly an important source of income for the people of Indian Sundarbans. In nineties, the most profitable livelihood in Indian Sundarbans was the culture of *Penaeus monodon*. However, due to white spot disease,

Table 4.5 An enterprise model for *Suaeda* and *Salicornia* farming

Model	Activities systems	Value creation	Value capture	Limitations
Global scenario	Business administration processes involving manpower with experience on halophyte farming; innovation and technology management considering the ecology of the halophytes and quality of this mangrove associate floral extracts; marketing and advertising (preferably digital)	Safety; scientific value; certifications, competitive prices, technological innovations in the domain of mangrove associate floral species	Economy of scale; large markets preferably in the domains of spin off products like fish feed, snacks and health drinks preparation	Limited intrinsic value creation due to standardization; market risks; natural disasters like wave actions, tsunami type massive wave surges, tidal surges, sea level rise, acidification, super-cyclones etc.; anthropogenic stress arising from aquaculture, tourism etc.
SMEs	Innovation and technology management considering the saline habitat and resource base (biomass); marketing and advertising; e-commerce	Customer oriented marketing innovations, services, organizational innovation emerging from mangrove associate-centric product development	Economy of niches/ segmentation	Difficulty with certification requirements; perception of benefits of sustainable practices; competitive disadvantages; lack of knowledge on carbon capture potential of *Suaeda* and *Salicornia*

(continued)

Table 4.5 (continued)

Model	Activities systems	Value creation	Value capture	Limitations
Traditional provisioning services	Sustainable management or extraction from vegetative parts of mangrove associate floral species; human consumption as salads, curries etc.	Trust in a community authority; customization benefits for local community (social responsibility)	Persuasive capacity for selling *Suaeda* and *Salicornia* based products and services preferably in the sectors of fish feed industry, snacks preparation units, health drinks preparation units etc.	Limited market size; low capacity; incapacity of complying with regulations (preferably related to conservation related acts and policies)
Hybrid innovative practitioners	Innovation and technology management considering the salinity preference of the mangrove associate species; specialization in provisioning services like raw materials for salt, fish feed, snacks, health drinks preparation etc.; marketing and advertising highlighting the carbon sequestration potential of the floral species (so that the carbon credit can also be obtained for halophyte farming); e-commerce	Customer oriented services & marketing innovations, partnerships with scientific and technological institutions and academic institutes competent in halophyte farming	Economy of niches; access to markets overseas	Logistic and regulatory limitations

Table 4.6 Strength of climate—resilient halophyte farming in the framework of Indian Sundarbans

Component	Western Sector	Central Sector	Eastern Sector
Availability of raw materials			
Survival and growth of halophytes (*Suaeda* and *Salicornia*)			
Congenial soil condition			
Availability of high saline estuarine water			

Strength Index

Maximum Strength	Moderate Strength	Less Strength

Table 4.7 Weakness of climate—resilient halophyte farming in the framework of Indian Sundarbans

Component	Western Sector	Central Sector	Eastern Sector
Awareness on halophyte farming			
Supply chain			
Institutional support			
Subsidy and bank loan			

Weakness Index

Maximum Weakness	Moderate Weakness	Less Weakness

Table 4.8 Opportunity of climate—resilient halophyte farming in the framework of Indian Sundarbans

Component	Western Sector	Central Sector	Eastern Sector
Innovative product development			
Fish feed, and food items development with high nutritional value			
Employment generation			
Non – conventional skill development			

Opportunity Index

Maximum Opporunity	Moderate Opportunity	Less Opportunity

Table 4.9 Threat on climate–resilient halophyte farming in the framework of Indian Sundarbans

Component	Western Sector	Central Sector	Eastern Sector
Erosion	Moderate Threat	Maximum Threat	Less Threat
Natural disaster	Maximum Threat	Maximum Threat	Maximum Threat
Salinity	Maximum Threat	Less Threat	Maximum Threat
Wave action	Moderate Threat	Maximum Threat	Less Threat

Threat Index

Maximum Threat	Moderate Threat	Less Threat

Table 4.10 (a) Fixed cost of *Suaeda* and *Salicornia* farming

Sl. No.	Particulars	Cost (Rs.)
1	Soil preparation for maintaining salinity and nutrients	3,00,000/-
2	Pump, electric motor, and electric lining	4,00,000/-
3	Drying and powdering equipment	5,50,000/-
4	Miscellaneous cost	1,50,000/-
Total fixed cost		14,00,000/-

Table 4.10 (b) Operational cost of *Suaeda* and *Salicornia* farming

Sl. No.	Particulars	Cost (Rs.)
1	Lease of the land in the supra-littoral zone in Indian Sundarbans	1,00,000/-
2	Chemical and organic fertilizer	1,00,000/-
3	Seed cost along with planting	2,00,000/-
4	Labour wages per annum @ Rs. 5,000 per head per month including harvesting cost (4 labours)	2,40,000/-
5	Fuel and electricity charges	1,20,000/-
6	Transportation charges	1,00,000/-
7	Miscellaneous charges	10,000/-
Total operational cost		8,70,000/-

Table 4.10 (c) Total cost of *Suaeda* and *Salicornia* farming

Sl. No.	Particulars	Cost (Rs.)
1	Fixed cost	14,00,000/-
2	Operational cost	8,70,000/-
3	Depreciation on fixed cost @10%/year	1,40,000/-
4	Interest on fixed investment @15%/year	2,10,000/-
Total cost		26,20,000/-

Table 4.10 (d) Selling price of *Suaeda* and *Salicornia*

Sl. No.	Particulars	SP (Rs.)
1	Sale of 1.2 tonnes @ Rs. 6,200 per kg	74,40,000/-
Net income (Total sale – Total cost)		48,20,000/ha/year

The selling price of the harvested, dried, and powdered halophytes has been calculated based on the standard price as stated in https://run.unl.pt/bitstream/10362/25208/1/Vieira_etall_2017.pdf

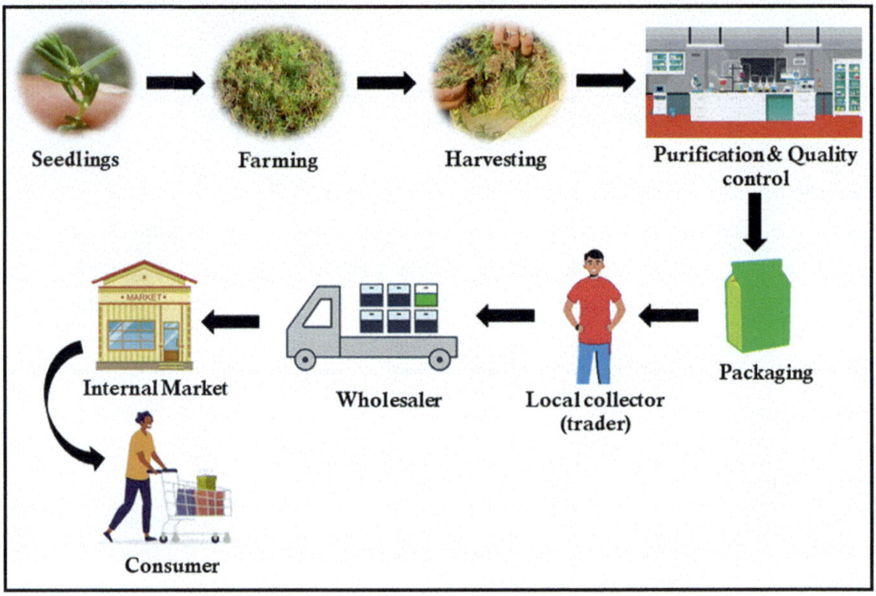

Fig. 4.16 Diagrammatic representation of supply chain in the halophyte farming sector

this sector of livelihood faced a great setback. The devastation in the shrimp culture sector in nineties was mainly due to mind set of the farmers to achieve short term gain without caring for the environment and feed quality. During those days, high stocking density of the prawn seed in the cultured pond and leftover protein—rich unused feed particles on the pond bottom deteriorated the water as well as the shrimp quality. White spot disease devasted the entire business and the farmers started to think for a new species. Based on this scenario, the culture of white shrimp *P. vannamei* was accepted by most of the farmers/shrimp culturists of Indian Sundarbans.

P. vannamei contains about 35% protein and its feed cost is less compared to *P. monodon feed*. The Feed Conversion Ratio (FCR) value is also less in *P. vannamei* culture (~1.2) compared to 1.6–1.8 in *P. monodon*. To make this venture sustainable, we undertook a pilot study in Bali Island (22° 04′ 35.17″ N and 88° 44′ 55.70″ E) during March–June 2023 to culture *P. vannamei* with *Salicornia*-based fish feed. The detail technology of our pilot scale project is stated here in a lucid way for clear understanding of the readers.

Our trial consisted of dietary treatment for two types of ponds, namely control and experimental. In the control pond, commercial feed available in the local market was provided and in the experimental pond, the *Salicornia*-based formulated feed was given throughout the culture periods of 4 months (March, 2023–June, 2023). Each pond was well connected to the adjacent estuary so that possible hydrological variations influence both the ponds simultaneously and uniformly.

The experimental feed was designed according to the nutritional requirements of shrimp (Mukhopadhyay et al. 2003). The floral ingredients (*Salicornia*) were incorporated at a level of 30% within the feed by reducing fish meal (Table 4.11). A control feed (with fish meal) was prepared as per the standard method by procuring ingredients from the local market. The feed ingredients were chosen based on their nutritional status, price, and year-round availability in the local market like Gosaba (this is the nearest island to our culture site with market facilities). The ingredients were weighed properly followed by a uniform mixture. The resulting mixture was steam cooked, cooled at room temperature, and finally pressed through a manual feed pelletizer. The pellets were dried in well aerated place under the shade for 2 days until it became sufficiently dry. Finally, they were packed and stored for further use. The proximate analysis of the formulated feeds were determined by following the standard methods like Kjeldhal for protein (Tecator 1987), Dubois for carbohydrate (Dubois et al. 1956), and Soxhlet for lipid (Folch et al. 1957). Flame photometric method was used to determine the levels of sodium and potassium.

S. brachiata has considerable protein and mineral contents due to which the species can be used as one of the main ingredients of shrimp feed (Table 4.12; Figs. 4.17 and 4.18).

The control feed purchased from the local market did not contain any floral ingredients and the proximate composition and the mineral contents of the same are highlighted in Figs. 4.19 and 4.20 respectively.

We analyzed the shrimp feed prepared from the dried leaves and stems of *S. brachiata* (experimental feed) and assessed their nutritional value. Although the

Table 4.11 Composition of shrimp feed for control and experimental ponds

Ingredients (in %)	Control pond	Experimental pond
Fish meal	30.0	25.0 (*Salicornia brachiata* leaf dried and powder)
Shrimp meal	5.0	5.0 (*Salicornia brachiata* stem dried and powder)
Soybean meal	5.0	10.0
Mustard oilcake	10.0	10.0
Sesame meal	10.0	10.0
Wheat bran	20.0	20.0
Rice bran	18.0	18.0
Oyster shell	1.0	1.0
Vitamin premix	1.0	1.0
Total	100.0	100.0

Table 4.12 Proximate composition of *S. brachiata* and *S. brachiata*-based formulated feed

Species and Product	Total protein (%)	Total carbohydrate (%)	Total fat (%)	Total fiber (%)	Ca (%)	Na (%)	K (%)
Salicornia brachiata	17.73 ± 1.89	57.12 ± 2.64	9.3 ± 0.15	4.66 ± 0.98	2.83 ± 0.19	4.14 ± 0.76	3.69 ± 0.63
Control feed	25.80 ± 1.98	31.47 ± 3.11	2.34 ± 0.84	8.12 ± 1.27	0.77 ± 0.05	0.21 ± 0.03	0.16 ± 0.04
Formulated feed	23.16 ± 2.02	61.66 ± 3.11	0.76 ± 0.09	7.83 ± 1.11	2.11 ± 0.11	3.34 ± 0.52	2.83 ± 0.44

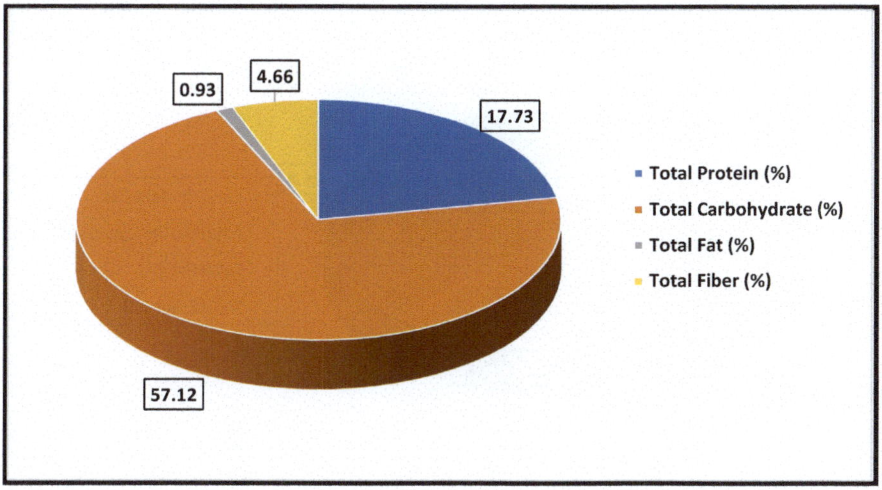

Fig. 4.17 Protein, carbohydrate, and fat content in *Salicornia brachiata*

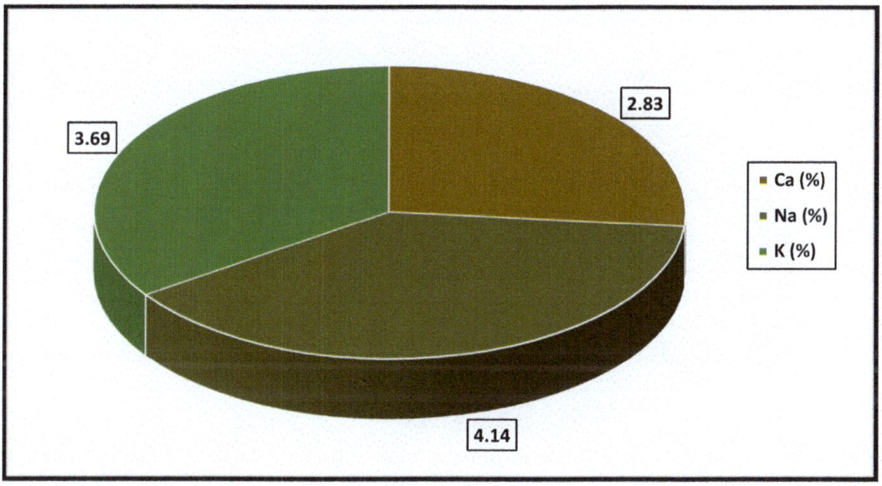

Fig. 4.18 Ca, Na, and K content in *Salicornia brachiata*

protein content decreased, but the levels of minerals were relatively high compared to the control feed (Figs. 4.21 and 4.22).

The feed prepared from the vegetative parts (leaf and stem) of the mangrove associate species also upgraded the water in terms of dissolved oxygen, surface water pH, nutrients, and soil organic carbon (Tables 4.13, 4.14, 4.15 and 4.16; Figs. 4.23, 4.24, 4.25, 4.26, 4.27, 4.28, 4.29, 4.30 and 4.31).

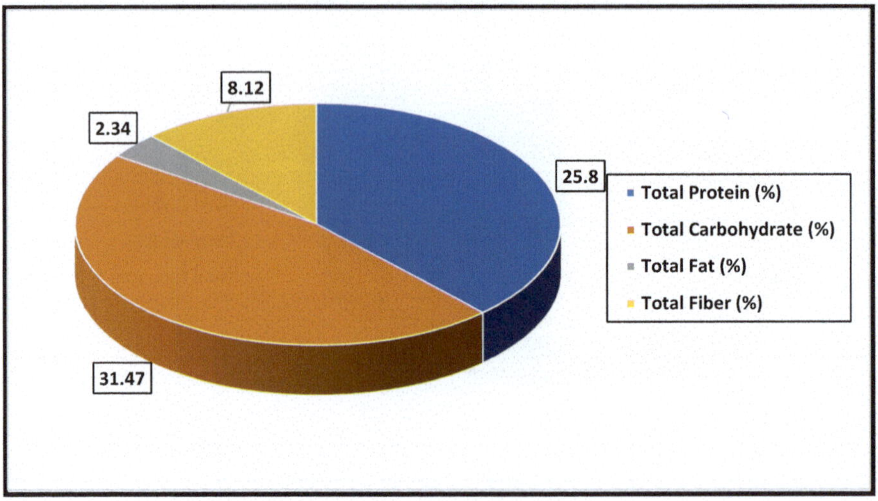

Fig. 4.19 Protein, carbohydrate, and fat content in control feed

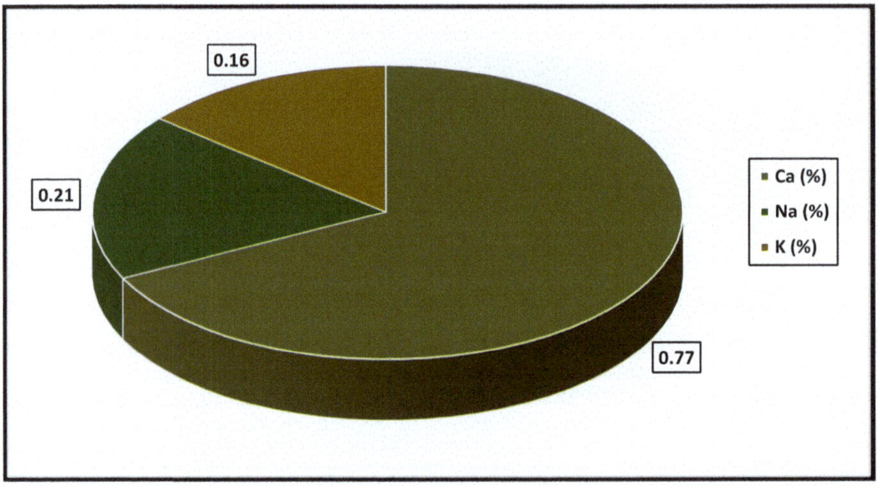

Fig. 4.20 Ca, Na, and K content in control feed

The production of shrimp was higher in the pond treated with *Salicornia*-based feed compared to the control. In our trial, the survival rate was 85% in the experimental pond, which is 16.44% higher than the control pond where the survival rate was 73%. The daily growth rate was high in the experimental pond (0.15 ± 0.002 g/day) compared to the control pond (0.11 ± 0.003 g/day). Interestingly, we observed the FCR value was much lower in the experimental pond (1.09), which is 18.35% lower than 1.29, the FCR value of the control pond. The lower value of FCR is an

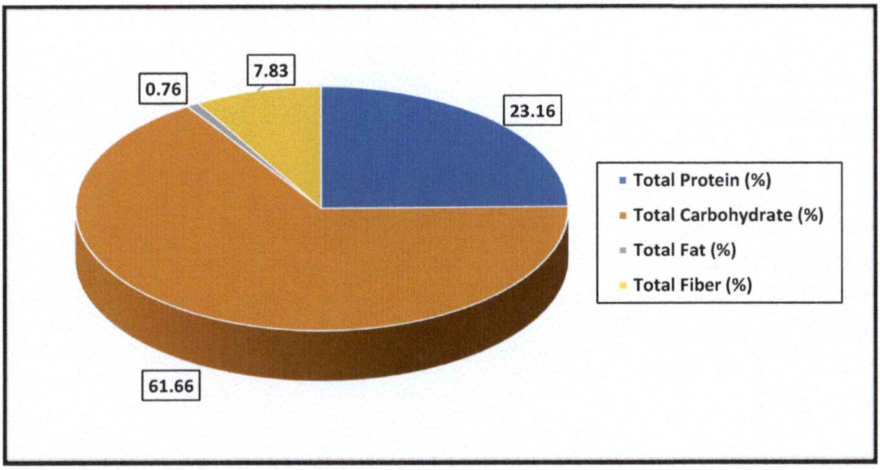

Fig. 4.21 Protein, carbohydrate, and fat content in experimental feed

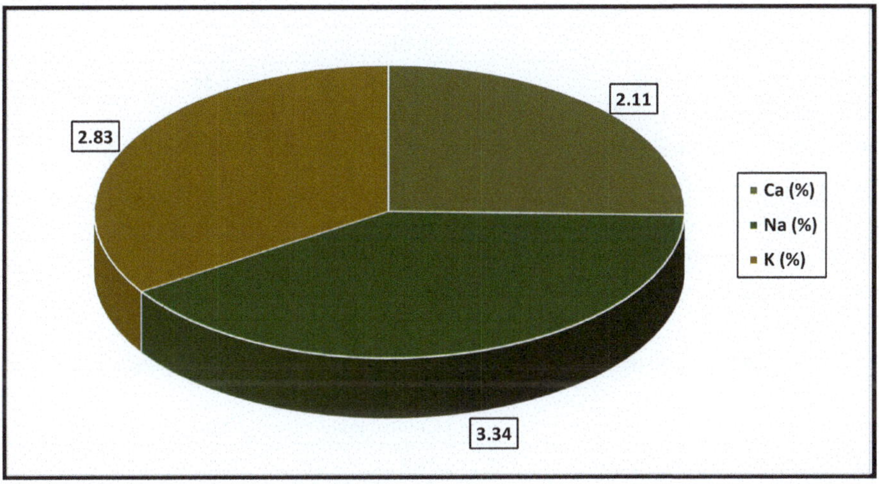

Fig. 4.22 Ca, Na, and K content in experimental feed

indication of lesser residual feed on the pond bottom, which otherwise may be the source of GHGs from unutilized feed.

We believe that people's participation/interest is the foundation of any initiative in a region. The people of Sundarbans are familiar with the fish feed (made from animal ingredients) available in the market that are manufactured by many multinational and Indian companies, and hence the concept of halophyte—based fish feed was

Table 4.13 Water and soil quality of control and experimental ponds in Indian Sundarbans during March, 2023

Parameters	Control pond	Experimental pond
Surface water temperature (°C)	30.1	29.9
Surface water pH	8.05	8.19
Surface water salinity (psu)	9.34	9.76
Dissolved oxygen (ppm)	5.18	5.93
Dissolved nitrate (μgm at l^{-1})	17.85	16.23
Dissolved phosphate (μgm at l^{-1})	2.98	2.73
Dissolved silicate (μgm at l^{-1})	65.83	61.29
Phytopigment (chlorophyll *a*) (mg m^{-3})	7.85	6.91
Soil organic carbon (%)	1.79	1.56

Table 4.14 Water and soil quality of control and experimental ponds in Indian Sundarbans during April, 2023

Parameters	Control pond	Experimental pond
Surface water temperature (°C)	32.1	31.9
Surface water pH	8.01	8.19
Surface water salinity (psu)	11.40	11.38
Dissolved oxygen (ppm)	5.04	5.97
Dissolved nitrate (μgm at l^{-1})	18.83	16.84
Dissolved phosphate (μgm at l^{-1})	3.02	2.69
Dissolved silicate (μgm at l^{-1})	65.87	61.36
Phytopigment (chlorophyll *a*) (mg m^{-3})	8.13	5.95
Soil organic carbon (%)	1.83	1.59

Table 4.15 Water and soil quality of control and experimental ponds in Indian Sundarbans during May, 2023

Parameters	Control pond	Experimental pond
Surface water temperature (°C)	34.8	33.9
Surface water pH	7.99	8.18
Surface water salinity (psu)	14.16	14.12
Dissolved oxygen (ppm)	4.96	5.91
Dissolved nitrate (μgm at l^{-1})	20.31	16.20
Dissolved phosphate (μgm at l^{-1})	3.16	2.64
Dissolved silicate (μgm at l^{-1})	65.71	61.29
Phytopigment (chlorophyll *a*) (mg m^{-3})	9.17	7.13
Soil organic carbon (%)	1.79	1.72

Table 4.16 Water and soil quality of control and experimental ponds in Indian Sundarbans during June, 2023

Parameters	Control pond	Experimental pond
Surface water temperature (°C)	36.7	36.1
Surface water pH	7.64	8.17
Surface water salinity (psu)	15.14	15.19
Dissolved oxygen (ppm)	4.91	5.96
Dissolved nitrate (μgm at l^{-1})	21.77	15.98
Dissolved phosphate (μgm at l^{-1})	3.53	2.67
Dissolved silicate (μgm at l^{-1})	65.81	61.54
Phytopigment (chlorophyll *a*) (mg m^{-3})	11.44	7.04
Soil organic carbon (%)	2.58	1.68

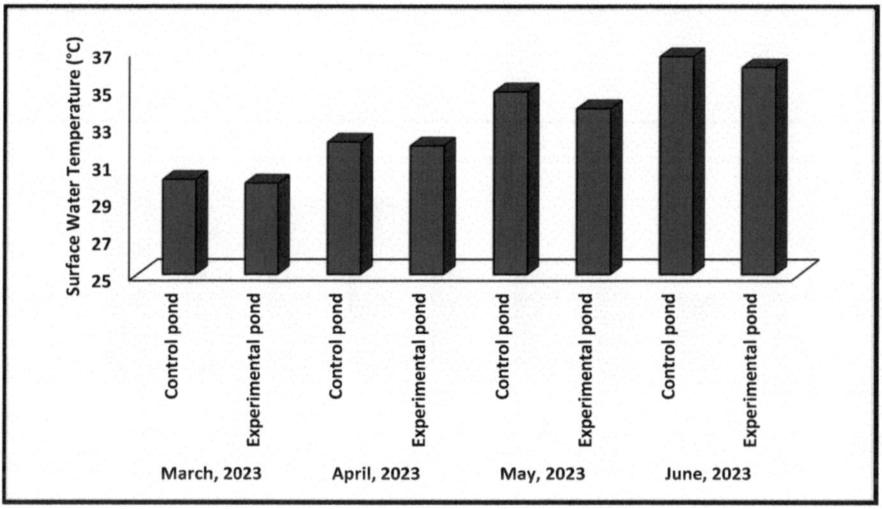

Fig. 4.23 Surface water temperature in the control and experimental ponds during the trial phase of 4 months

attempted to establish through SWOT analysis after a detailed group discussion with the local shrimp culturists (n = 210), Panchayat members (n = 45), agriculturists (n = 134) and local NGOs (n = 7). The results are depicted in Tables 4.17, 4.18, 4.19 and 4.20.

Tables 4.21, 4.22, 4.23 and 4.24 present the Benefit Cost Analysis of the climate—resilient *Salicornia*-based shrimp feed.

A plant—based (herbal) shrimp feed supply chain can be described as a set of independent farmers (both for farming *Salicornia* and shrimp), processors, distributors, and wholesalers/retailers etc. who work together to supply the organic feed for the shrimp species (*Penaeus vannamei*) or derived product to the consumer. The

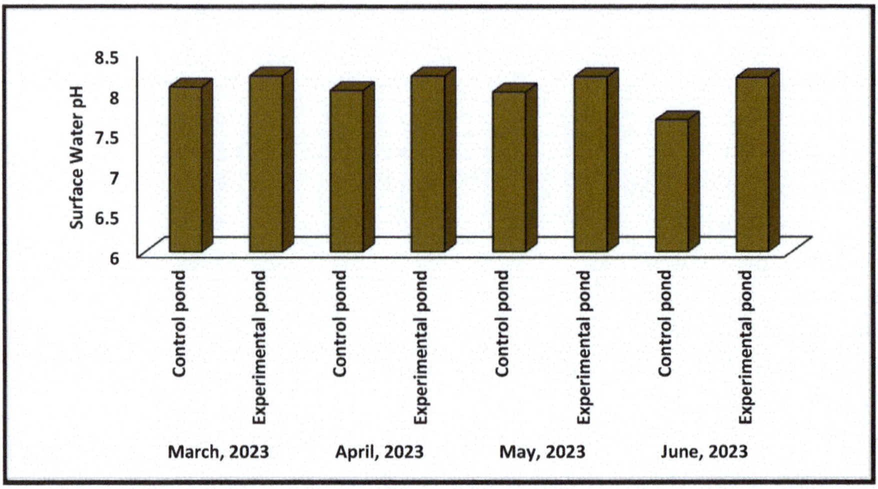

Fig. 4.24 Surface water pH in the control and experimental ponds during the trial phase of 4 months

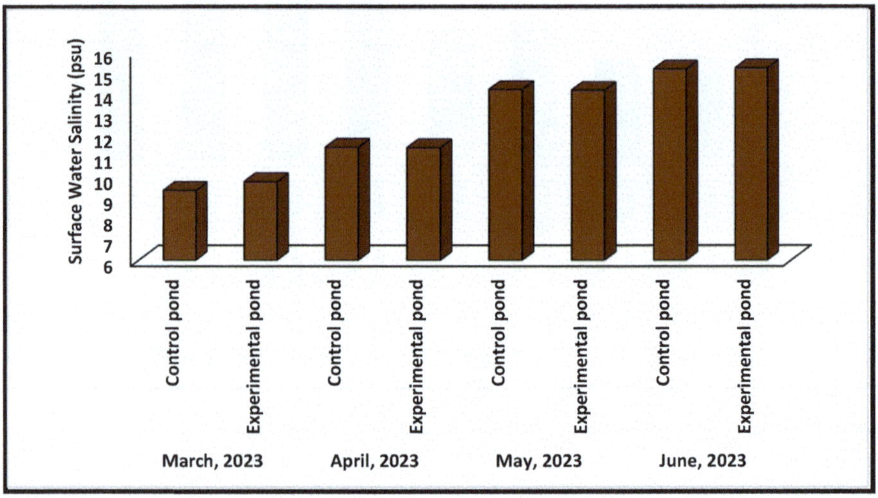

Fig. 4.25 Surface water salinity in the control and experimental ponds during the trial phase of 4 months

increasing trend of seafood consumption has triggered the market demand for shrimp feed. Furthermore, the Government of India has given high weightage to organic shrimp farming to ensure sustainability to the environment. On this background, the *Salicornia*-based feed of *Penaeus vannamei* seems to have a fruitful market with the supply chain as shown in Fig. 4.32.

The high saline zone in Indian Sundarban is restricted in the central sector where islands like Jharkhali, Bali Island, Chotomollakhali, Kumirmari etc. are located. The

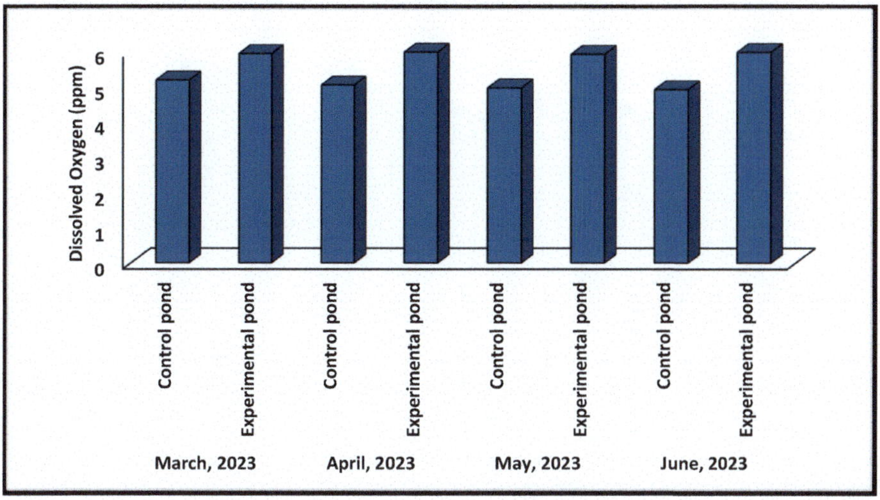

Fig. 4.26 Dissolved Oxygen (DO) in the control and experimental ponds during the trial phase of 4 months

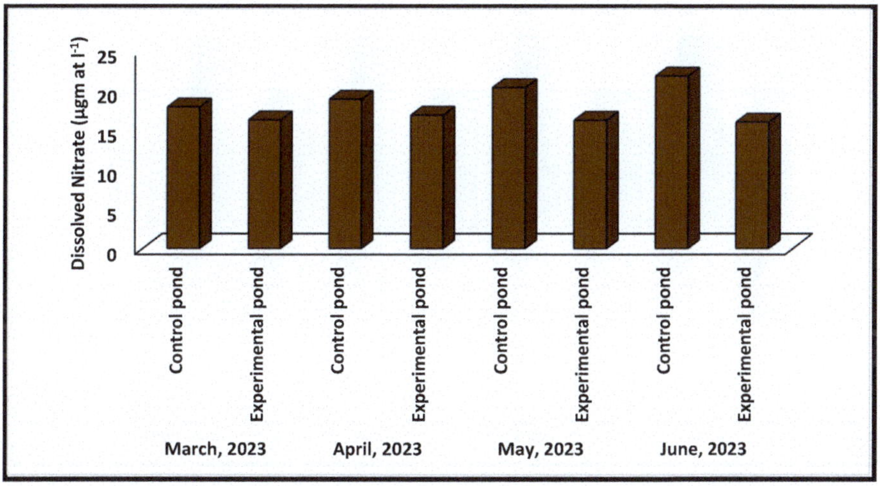

Fig. 4.27 Dissolved nitrate in the control and experimental ponds during the trial phase of 4 months

supra-littoral habitat of the high saline zone is the rich survival ground of mangrove associate species like *Suaeda maritima* and *Salicornia brachiata*, which can open a new dimension of livelihood in Indian Sundarbans, provided awareness on the nutritional values of these species are percolated to people of the region and adequate institutional support are provided. However, salinity is the primary driver for halophyte farming and other spin-off products from halophytes, which is very congenial in the central sector of Indian Sundarbans. Thus, to sum-up we can conclude that if

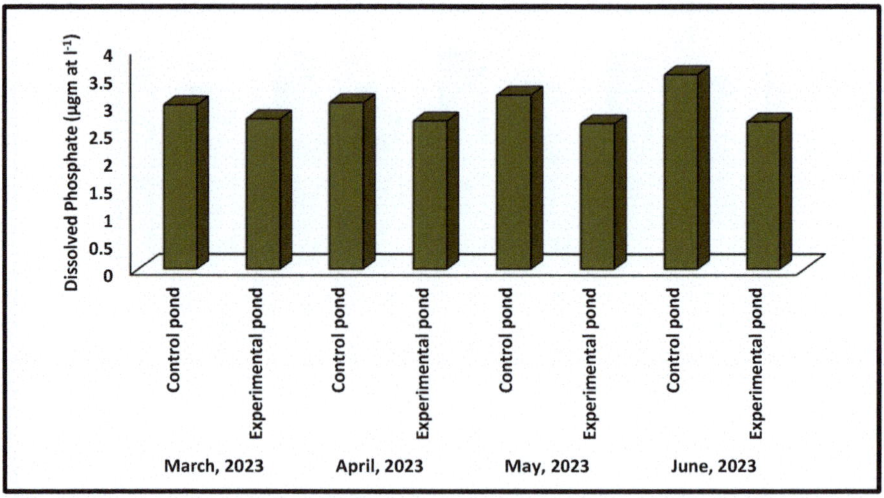

Fig. 4.28 Dissolved phosphate in the control and experimental ponds during the trial phase of 4 months

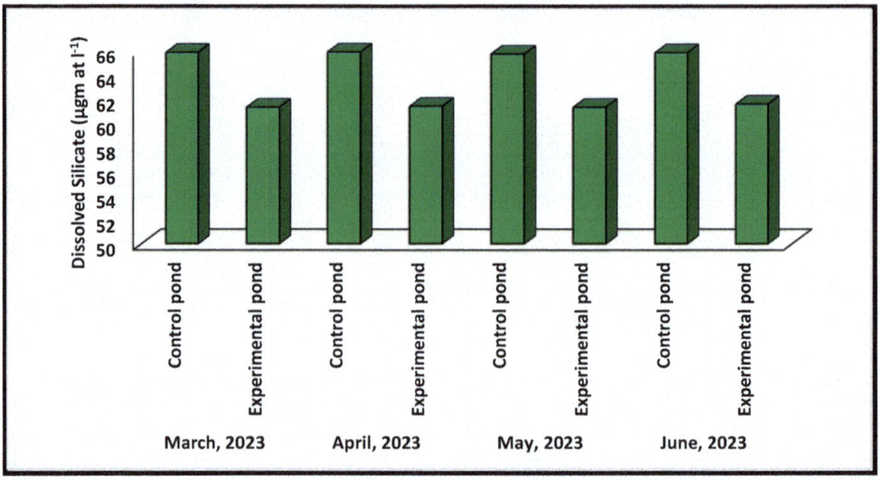

Fig. 4.29 Dissolved silicate in the control and experimental ponds during the trial phase of 4 months

institutional support is extended in terms of skill development, finance, and disaster management, then halophyte farming can provide a new horizon in the livelihood sector of central Indian Sundarbans. These halophytes, which are basically mangrove associate species, can be the source materials for fish feed, food products, cosmetics, and several pharmaceutical compounds.

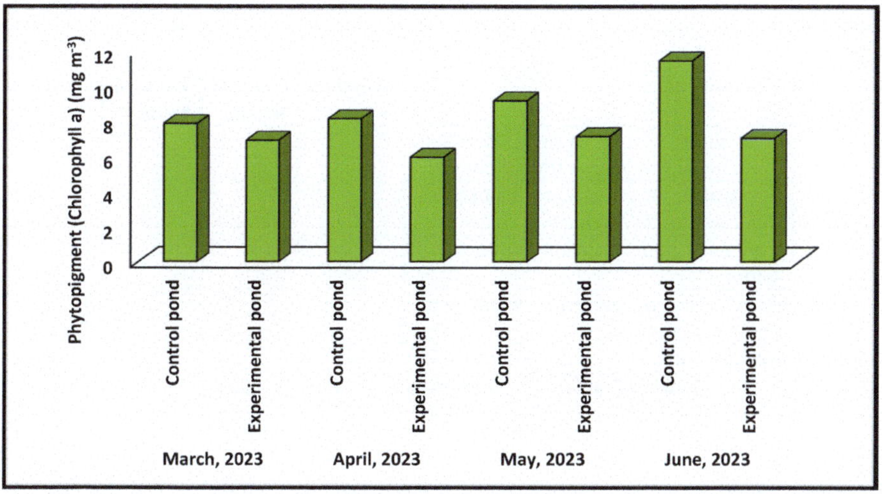

Fig. 4.30 Phytopigment level in the control and experimental ponds during the trial phase of 4 months

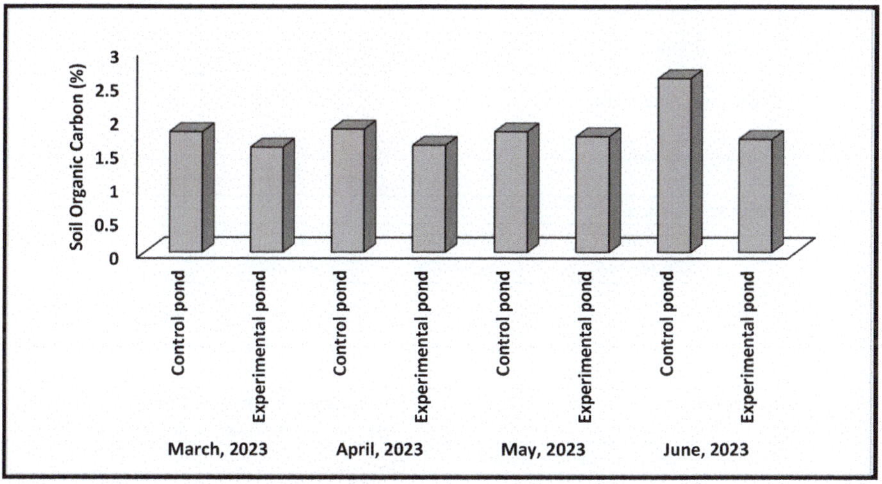

Fig. 4.31 Soil Organic Carbon (SOC) in the control and experimental ponds during the trial phase of 4 months

4.2 Alternative Livelihoods for Medium Saline Zone

In the medium saline zone of Indian Sundarbans, livelihoods like oyster culture and seaweed culture can be initiated, but the hindrances in launching such projects are the complete ignorance of the local people about the species and their uses, lack of institutional support, water quality in terms of salinity, pH and pollution level that are

Table 4.17 Strength of climate—resilient *Salicornia*-based shrimp feed in the framework of Indian Sundarbans

Component	Western Sector	Central Sector	Eastern Sector
Availability of raw materials (*Salicornia brachiata*) from the wild			
Survival and growth of halophytes (*Suaeda* and *Salicornia*)			
Congenial soil condition (in terms of salinity)			
Availability of estuarine water with optimum salinity			

Strength Index

Maximum Strength	Moderate Strength	Less Strength

Table 4.18 Weakness of climate—resilient *Salicornia*-based shrimp feed in the framework of Indian Sundarbans

Component	Western Sector	Central Sector	Eastern Sector
Awareness on *Salicornia*-based shrimp feed			
Supply chain			
Institutional support			
Subsidy and bank loan			

Weakness Index

Maximum Weakness	Moderate Weakness	Less Weakness

Table 4.19 Opportunity of climate—resilient climate—resilient *Salicornia*-based shrimp feed in the framework of Indian Sundarbans

Component	Western Sector	Central Sector	Eastern Sector
Innovative product development			
Fish feed, and food items development with high nutritional value			
Employment generation			
Non – conventional skill development			

Opportunity Index

Maximum Opporunity	Moderate Opportunity	Less Opportunity

Table 4.20 Threat on climate—resilient *Salicornia*-based shrimp feed in the framework of Indian Sundarbans

Component	**Western Sector**	**Central Sector**	**Eastern Sector**
Erosion			
Natural disaster			
Salinity			
Wave action			

Threat Index

Maximum Threat	**Moderate Threat**	**Less Threat**

Table 4.21 Fixed cost of *Salicornia*-based shrimp feed (calculation based on 10 tonnes production/month)

Sl. No.	Particulars	Cost (Rs.)
1.	Pump, electric motor, and electric lining	2,00,000/-
2.	Drying, grinding and pelletizer equipment	1,00,000/-
3.	Miscellaneous cost	20,000/-
Total fixed cost		3,20,000/-

Table 4.22 Operational cost of *Salicornia*-based shrimp feed

Sl. No.	Particulars	Cost (Rs.)
1	*Salicornia* farming including labour charges	1,00,000/-
2	Feed ingredients (Other than dried *Salicornia*)	40,000/-
3	*Salicornia* dust (dried)	60,000/-
4	Labour wages per annum @ rate of Rs.5000 per month for three labours	1,80,000/-
5	Fuel and electricity charges	50,000/-
6	Miscellaneous charges	20,000/-
Total operational cost		4,50,000/-

Table 4.23 Total cost of *Salicornia*-based shrimp feed

Sl. No.	Particulars	Cost (Rs.)
1	Fixed cost	3,20,000/-
2	Operational cost	4,50,000/-
3	Depreciation on fixed cost @10%/year	32,000/-
4	Interest on fixed investment @15%/year	48,000/-
Total cost		8,50,000/-

Table 4.24 Selling price of *Salicornia*-based shrimp feed

Sl. No.	Particulars	SP (Rs.)
1	Sale of 10 tonnes @ Rs. 120 per kg	12,00,000/-
Net income (Total sale – Total cost)		3,50,000/-

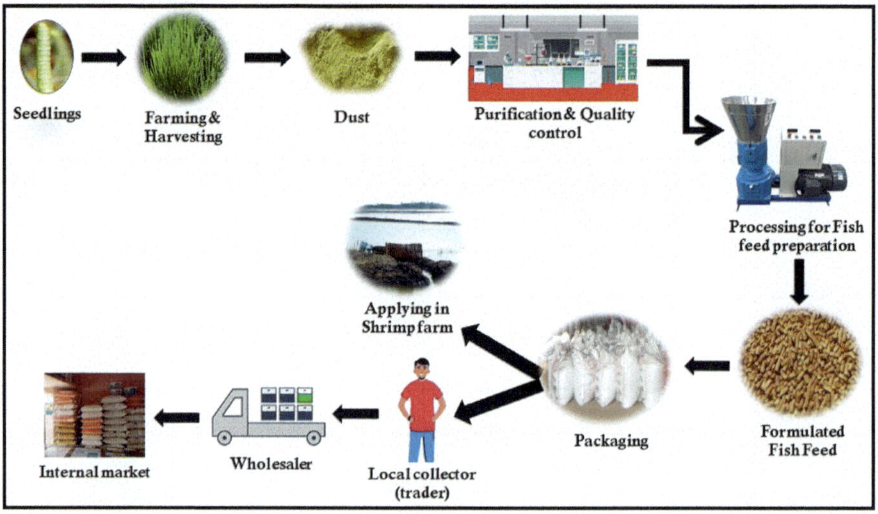

Fig. 4.32 Diagrammatic representation of supply chain in the *Salicornia*-based shrimp feed sector

highly pronounced in the western Indian Sundarbans. The region is highly urbanized with large scale fish landing activities, tourism, pollution from the adjacent Haldia industrial belt etc. The cultured products may be contaminated with heavy metals, pesticides etc. (bioaccumulation) and may lose their export values. In addition to these, the turbid waters of Indian Sundarban estuaries often limit the growth of seaweeds and phytoplankton that are the primary food of oysters. The central sector of Indian Sundarbans is ideal for initiating oysters and seaweed culture through local level befitted technology/methods that primarily involves screening the silt particles and proper water quality management.

4.2.1 Oyster Culture

Indian Sundarbans is the survival ground of various types of molluscs consisting of mainly bivalves and gastropods (Banerjee et al. 2015). Oysters are abundantly available in the mid-saline and even in the high-saline central sectors of the delta region. However, they cannot thrive below a salinity level of 10.0 psu. Three species of oysters are commonly seen growing naturally in the Sundarban mangrove belt

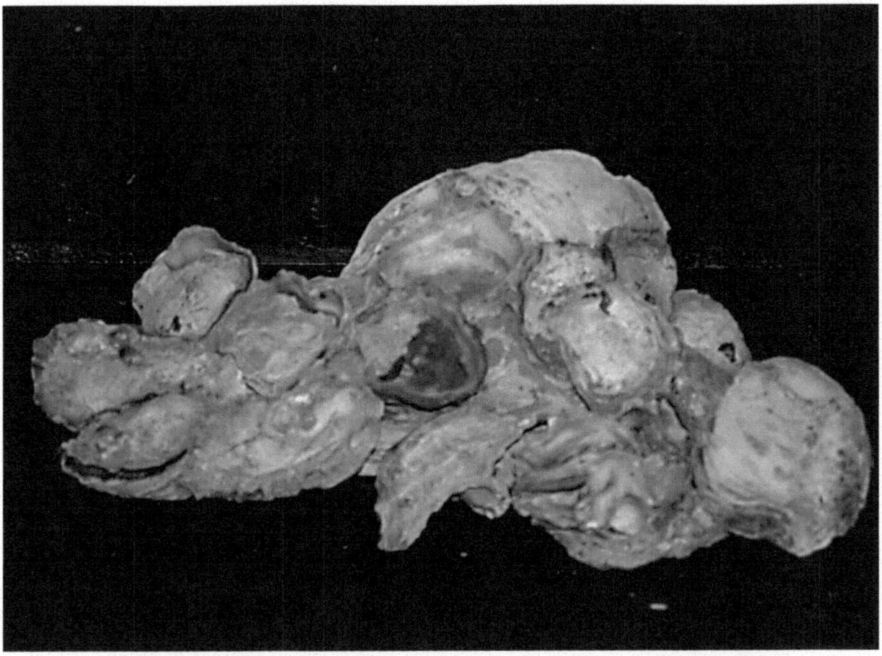

Fig. 4.33 *Saccostrea cucullata*—the species is highly variable in shape, growing in clusters on rocks, bricks, wooden piles, or jetties throughout the Indian Sundarbans (except in the low saline belt)

namely *Saccostrea cucullata, Crassostrea cuttackensis* and *Crassostrea madrasensis* (Figs. 4.33, 4.34 and 4.35). They are found attached on sluice gates, mangrove trunk, jetties, and hard structures associated with fish landing stations.

There are two common methods for culturing oysters and these are (i) On-bottom culture and (ii) Off-bottom culture. In case of Indian Sundarbans, considering the background of siltation rate and turbidity of the aquatic phase, off-bottom culture is most preferable and among various types of off-bottom culture most economic methods are Rack and string culture (Fig. 4.36) and Rack and tray culture (Fig. 4.37).

We have made detailed analysis on the feasibility of oyster culture in Indian Sundarbans and pinpointed few features that can accelerate the growth of oysters in the present geographical locale.

1. Protection against excessive wind and wave action (particularly applicable for off-bottom culture).
2. Monitoring and maintenance of water quality in terms of sewage and industrial wastes (being sedentary in nature oysters tend to accumulate conservative substances in the body tissues).
3. Water quality in terms of salinity, pH and temperature (optimum salinity requirement for culture is 18–32 psu, pH requirement is around 8.30 and temperature requirement is 25–30 °C).

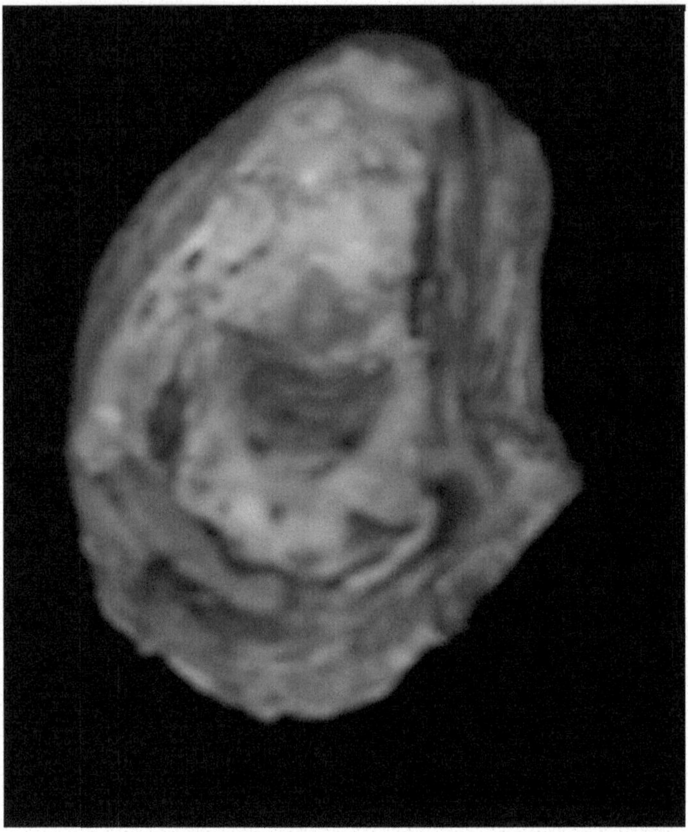

Fig. 4.34 *Crassostrea gryphoides*—the species is mainly found on the lower height of sluice gates and sometimes on hard substrata like lighthouse, boulders etc.

4. Tidal characteristics and current pattern (as frequent water exchange is extremely important for oyster culture).
5. Optimum nutrient level and adequate phytoplankton density (as nutrition of the oyster species is obtained through consumption of phytoplankton).
6. Presence of predators, like boring gastropod *Cymatium*, starfish, crabs etc. (sites with high population and diversity of predators are always avoided as they cause massive damage to oyster population).
7. Availability and abundance of adult oyster stock in the vicinity of the culture site (this is for ensuring oyster seed supply on regular basis).

We also conducted a 2-day workshop on 22nd and 23rd March, 2023 on the feasibility of oyster culture involving the residents of Sundarbans and felt their pulses based on SWOT analysis (Tables 4.25, 4.26, 4.27 and 4.28).

Tables 4.29, 4.30, 4.31 and 4.32 present the Benefit Cost Analysis of the climate—resilient Oyster culture in the mid saline zone of Indian Sundarbans.

Fig. 4.35 *Crassostrea madrasensis*—the species is observed in the lower stretch of Indian Sundarbans, attached to the sluice gates, bricks and dykes and to the lighthouse and jetties

The supply chain of oysters encompasses oyster farmers, traders, wholesalers and end consumers like hotels, restaurants, malls etc. (Fig. 4.38). The Central Marine Fisheries Research Institute (CMFRI) directly markets live oysters in high-end restaurants. Similarly, several Self-Help Groups (SHG) also sell oysters to high class hotels and restaurants.

4.2.2 Seaweed Culture

Seaweeds are abundantly available in the estuarine water of Indian Sundarbans, but the most dominant species are *Enteromorpha intestinalis* and *Ulva lactuca* (Figs. 4.39 and 4.40). They can tolerate medium salinity, but the main hindrances to seaweed

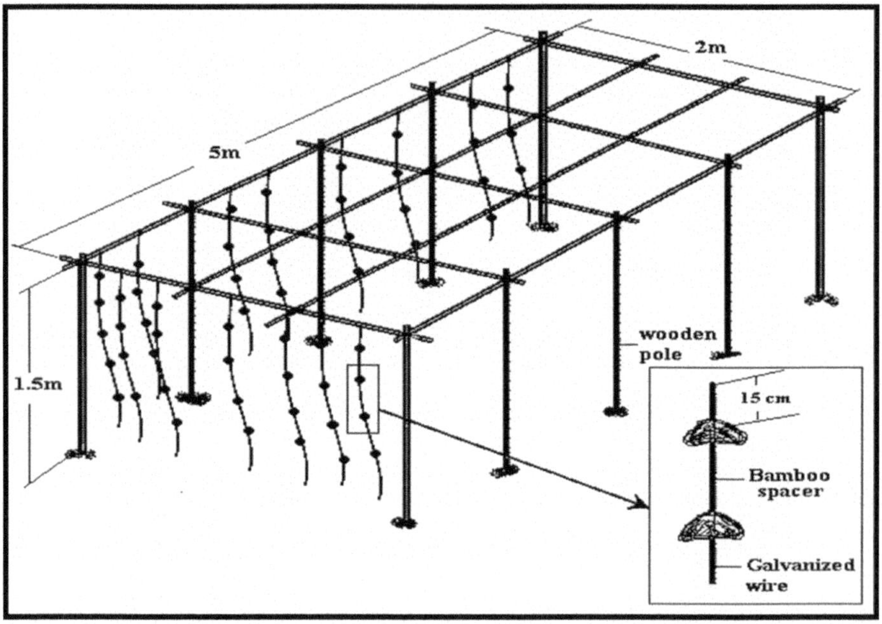

Fig. 4.36 Rack and string culture for oyster farming

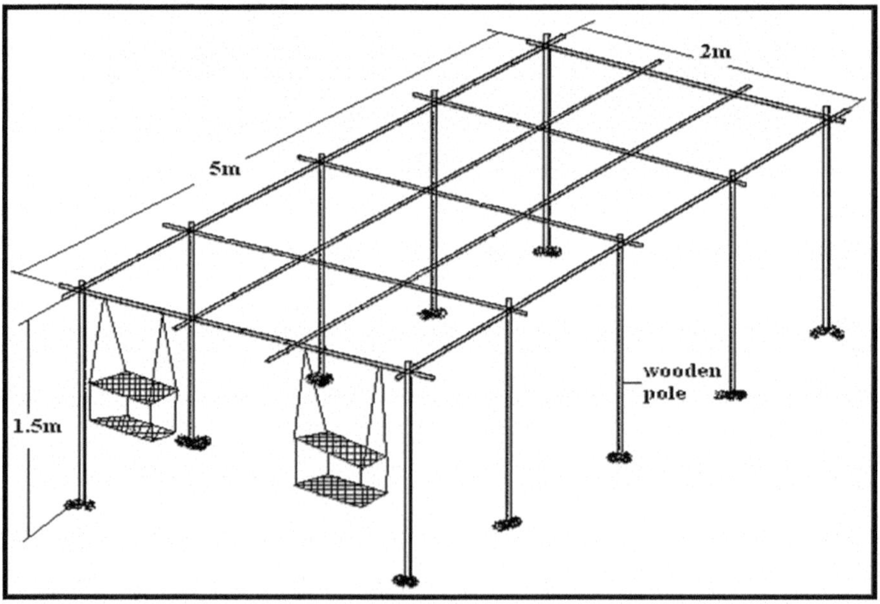

Fig. 4.37 Rack and tray culture for oyster farming

Table 4.25 Strength of climate—resilient oyster culture in the framework of Indian Sundarbans

Component	Western Sector	Central Sector	Eastern Sector
Awareness of the local people	Less Strength	Moderate Strength	Maximum Strength
Abundance of larvae in the estuarine water that can lead to natural spat on hard substrata	Moderate Strength	Maximum Strength	Moderate Strength
Congenial aquatic salinity preferably between 10-25 psu	Less Strength	Maximum Strength	Less Strength
Abundance of phytoplankton in the estuarine water that serve as the natural feed of oyster	Less Strength	Moderate Strength	Maximum Strength

Strength Index

Maximum Strength	Moderate Strength	Less Strength

Table 4.26 Weakness of climate—resilient oyster culture in the framework of Indian Sundarbans

Component	Western Sector	Central Sector	Eastern Sector
Awareness of the local people	Maximum Weakness	Maximum Weakness	Maximum Weakness
Institutional support	Maximum Weakness	Maximum Weakness	Maximum Weakness
Subsidy and bank loan	Maximum Weakness	Maximum Weakness	Maximum Weakness
Supply chain for oyster	Moderate Weakness	Maximum Weakness	Maximum Weakness

Weakness Index

Maximum Weakness	Moderate Weakness	Less Weakness

Table 4.27 Opportunity of climate—resilient oyster culture in the framework of Indian Sundarbans

Component	Western Sector	Central Sector	Eastern Sector
Employment generation	Moderate Opportunity	Maximum Opporunity	Less Opportunity
R & D establishment	Maximum Opporunity	Moderate Opportunity	Less Opportunity
Nutritional value	Moderate Opportunity	Maximum Opporunity	Maximum Opporunity
Foreign exchange earning	Maximum Opporunity	Maximum Opporunity	Less Opportunity

Opportunity Index

Maximum Opporunity	Moderate Opportunity	Less Opportunity

Table 4.28 Threat on climate—resilient climate—resilient oyster culture in the framework of Indian Sundarbans

Component	**Western Sector**	**Central Sector**	**Eastern Sector**
Natural disaster			
Acidification			
Dilution factor			
Pollution			

Threat Index

Maximum Threat	**Moderate Threat**	**Less Threat**

Table 4.29 Fixed cost of oyster farming (farm dimension 100 m × 100 m)

Sl. No.	Particulars	Amount (Rs.)
1	Bamboo poles—3,000 Nos. @ Rs. 300/-	9,00,000/-
2	Rope (for ren making + farm construction)—1,000 kg @ Rs. 250/- per kg	2,50,000/-
3	Miscellaneous cost	1,00,000/-
Total fixed cost		12,50,000/-

Table 4.30 Operational cost of oyster farming

Sl. No.	Particulars	Amount (Rs.)
1	Shell—6,00,000 Nos. @ Rs. 1/-	6,00,000/-
2	Ren making charges—1,20,000 Nos. @ Rs. 3/-	3,60,000/-
3	Farm construction charges	1,50,000/-
4	Installation of spat settler	1,00,000/-
5	Harvesting charges	2,00,000/-
6	Canoe hiring charges	50,000/-
7	Depuration charges	50,000/-
8	Shell on (Single oyster declumping)	2,00,000/-
9	Fuel charges (LPG)	1,00,000/-
10	Shucking charges	2,00,000/-
Total operational cost		20,10,000/-

Table 4.31 Total cost of oyster farming

Sl. No.	Particulars	Amount (Rs.)
1	Fixed cost	12,50,000/-
2	Operational cost	20,10,000/-
3	Depreciation on fixed cost @10%/year	1,25,000/-
4	Interest on fixed investment @15%/year	1,87,500/-
Total cost		35,72,500/-

Table 4.32 Selling price of oyster

Sl. No.	Particulars	Amount (Rs.)
1	Oyster—12,500 kg @ Rs. 380 per kg	47,50,000/-
Net iincome (Total sale − Total cost)		11,77,500/-

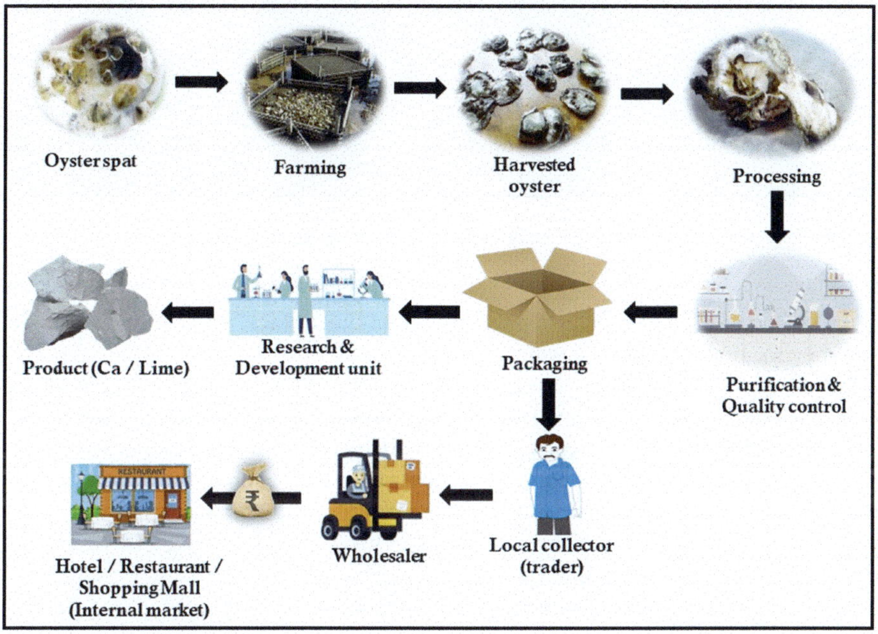

Fig. 4.38 Diagrammatic representation of supply chain in the oyster culture sector

culture in the present site are the siltation and contaminated water, particularly in the western Indian Sundarbans (Pramanick et al. 2015, 2016; Amin et al. 2018; Agarwal et al. 2022). In Indian Sundarbans, *Enteromorpha* can be cultivated through pond farming. Seed material/thallus introduced tied with long line ropes and nets can be cultured in the brackish water system at sub-surface level. The water depth should be maintained between 45–50 cm. Fertilization needs to be carried out with organic

Fig. 4.39 *Enteromorpha intestinalis*, major seaweed species in Indian Sundarbans

and inorganic fertilizers to boost up the growth of the species. The entire culture site should be located near the sea water.

The people of Sundarbans are not at all familiar with the seaweed culture and the provisional services obtained from the seaweed species and hence the concept of seaweed culture was percolated to the local inhabitants after a series of workshops during 24th March to 30th March, 2023. At the end, a SWOT analysis was carried out after a detailed group discussion with the local shrimp culturists (n = 188), panchayat members (n = 22), agriculturists (n = 166) and local NGOs (n = 4). The results are depicted in Tables 4.33, 4.34, 4.35 and 4.36.

We attempted a seaweed culture in a very crude way by tying thalli of *Enteromorpha intestinalis* (green seaweed under Chlorophyceae) ropes at Chemaguri (21° 39′ 58.15″ N and 88° 10′ 07.03″ E) creek during February, 2023 and within a period of 4 months, there was 80% increase in biomass (Fig. 4.41), which proves the suitability of the site for adopting seaweed culture as climate resilient alternative livelihood.

Tables 4.37, 4.38, 4.39 and 4.40 present the Benefit Cost Analysis of the climate—resilient seaweed culture in the mid saline zone of Indian Sundarbans.

The distribution of seaweed from the production center (seaweed farming unit in the neritic zone) to end consumers involves farmers, processors, traders, wholesalers, and consumers who purchase them from the shopping mall either in fresh or dried form as shown in the marketing channel in Fig. 4.42.

Fig. 4.40 *Ulva lactuca*—abundantly available in Indian Sundarbans

Table 4.33 Strength of climate—resilient seaweed culture in the framework of Indian Sundarbans

Component	Western Sector	Central Sector	Eastern Sector
Awareness of the local people			
Abundance of *Enteromorpha* sp and *Ulva* sp that are of great demand in the food, cosmetic and in fish feed preparation			
Aquatic salinity			
Nutrient levels in the estuarine water			

Strength Index

Maximum Strength	Moderate Strength	Less Strength

4.3 Low Saline—Based Alternative Livelihoods

In Indian Sundarbans, there are low saline belts in the western and eastern sectors, where the salinity in the monsoon goes down to even zero psu. There are also several pockets/canals where rain water accumulates and salinity drops down below 2 psu. These areas are suitable for the survival and growth of *Sonneratia* spp. and *Azolla*, which is a floating fern symbiotically associated with a nitrogen fixing bacteria known as *Anabaenae azollae*.

Table 4.34 Weakness of climate—resilient seaweed culture in the framework of Indian Sundarbans

Component	Western Sector	Central Sector	Eastern Sector
Awareness of the local people			
Institutional support			
Subsidy and bank loan			
Supply chain for seaweed			

Weakness Index

Maximum Weakness	Moderate Weakness	Less Weakness

Table 4.35 Opportunity of climate—resilient seaweed culture in the framework of Indian Sundarbans

Component	Western Sector	Central Sector	Eastern Sector
Employment generation			
R & D establishment			
Nutritional value			
Foreign exchange earning			

Opportunity Index

Maximum Opporunity	Moderate Opportunity	Less Opportunity

Table 4.36 Threat on climate—resilient seaweed culture in the framework of Indian Sundarbans

Component	Western Sector	Central Sector	Eastern Sector
Erosion and siltation			
Natural disaster			
Salinity			
Pollution			

Threat Index

Maximum Threat	Moderate Threat	Less Threat

Fig. 4.41 Trial of seaweed culture in the Chemaguri creek during February 2023–May 2023

Table 4.37 Fixed cost of seaweed farming

Sl. No.	Particulars	Amount (Rs.)
1	Five-toothed iron anchor—4 pcs. (each 100 kg) @ Rs. 110 per kg	44,000/-
2	Nylon ropes—100 kg @ Rs. 150 per kg	15,000/-
3	Iron Pin type Anchor Bolt—2000 pcs. @ Rs. 1.50 per pc	3,000/-
4	Seedlings/thallus—500 kg @ Rs. 5 per kg	2,500/-
5	HDPE net tying rope—10 kg @ Rs. 200 per kg	2,000/-
6	Anchoring rope—10 kg @ Rs. 200 per kg	2,000/-
7	Miscellaneous cost	50,000/-
Total fixed cost		1,18,500/-

Table 4.38 Operational cost of seaweed farming

Sl. No.	Particulars	Amount (Rs.)
1	Nitrogenous fertilizer—100 kg @ Rs. 70 per kg	7,000/-
2	Labour charges—2 heads @ Rs. 5,000 per head per month for 9 months	90,000/-
3	Transportation	50,000/-
4	Miscellaneous	10,000/-
Total operational cost		1,57,000/-

Table 4.39 Total cost of seaweed farming

Sl. No.	Particulars	Amount (Rs.)
1	Fixed cost	1,18,500/-
2	Operational cost	1,57,000/-
3	Depreciation on fixed cost @10%/year	11,850/-
4	Interest on fixed investment @15%/year	17,775/-
Total cost		3,05,125/-

Table 4.40 Selling price of seaweed

Sl. No.	Particulars	Amount (Rs.)
1	Seaweed—1500 kg @ Rs. 350 per kg	5,25,000/-
Net income (Total sale – Total cost)		2,19,875/-

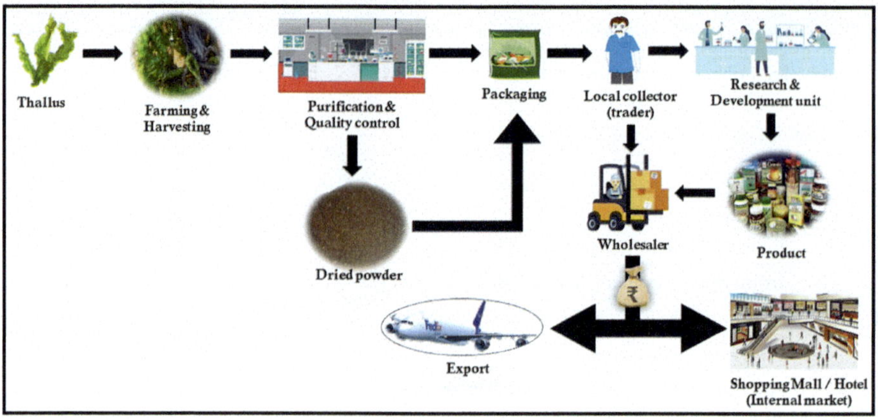

Fig. 4.42 Diagrammatic representation of supply chain in the seaweed farming sector

4.3.1 *Health Drink from* Sonneratia Caseolaris

Sonneratia caseolaris is a freshwater loving mangrove that grows luxuriantly in the western and eastern sectors of Indian Sundarbans. The species is also available in completely freshwater regions like Konnagar (22° 42′ 26.70″ N and 88° 21′ 27.50″ E) and Budge Budge (22° 28′ 18.00″ N and 88° 08′ 15.50″ E) adjacent to the city of Kolkata along the River Ganges in the state of West Bengal (Fig. 4.43). The fruit of the species has high nutritional value with considerable concentration of vitamin C (Figs. 4.44 and 4.45). Considering its abundance in the low saline region of Indian Sundarbans, innovative alternative livelihood has been proposed by us to promote both economic and health security. We undertook a pilot project to prepare health drinks from the fruit pulp of *S. caseolaris* after procuring the samples from Konnagar area in the Hooghly district of West Bengal.

Fig. 4.43 *Sonneratia caseolaris* along the Bank of the River Hooghly in Budge Budge in the state of West Bengal

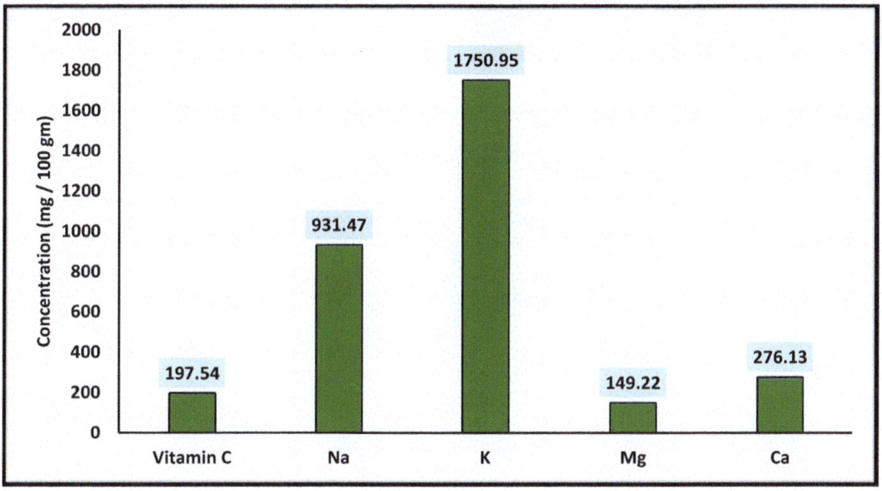

Fig. 4.44 Concentration of Vitamin C along with other major elements in *S. caseolaris* fruit pulp

The flow chart for preparing the health drink is provided in Fig. 4.46.

The prepared health drink was also analyzed to evaluate the quality of the product and we observed satisfactory result from the point of view of human consumption (Table 4.41).

This mangrove fruit derived health drink will not only upgrade the economic profile of the local people, but will also enhance the immunity power of the consumer. The health drink rich in vitamin C has the possibility of providing several health benefits as listed here (Table 4.42).

It is no exaggeration to say that the victim of COVID was minimum in this delta complex, which might be related to the high immunity power of the islanders that is obtained from consumption of this raw fruit on regular basis, although complete isolation of the island clusters from the main land (like the city of Kolkata) is also a major factor (Mitra 2023). We therefore undertook respondent analysis to feel the pulse of the local people for their involvement in such uncommon non-conventional livelihood (Table 4.43).

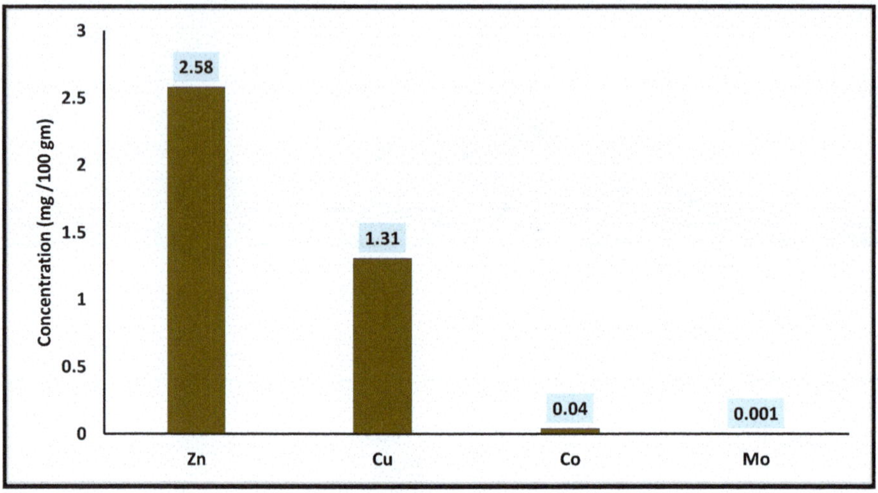

Fig. 4.45 Concentration of minor elements in *S. caseolaris* fruit pulp

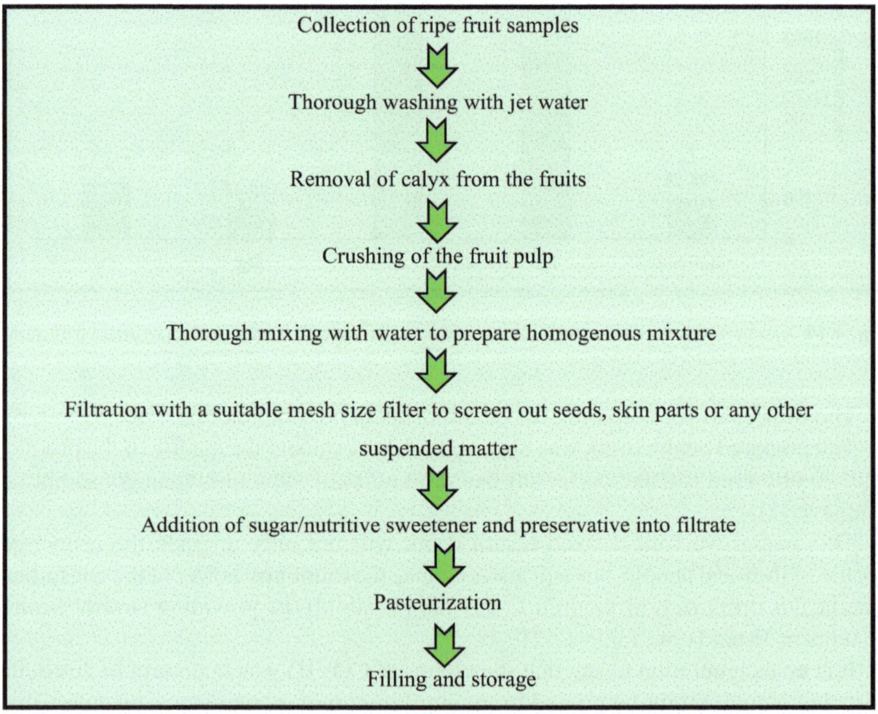

Fig. 4.46 Schematic diagram of preparation of health drink from *S. caseolaris* fruit

Table 4.41 Nutritional value of health drink from *S. caseolaris* fruit pulp

	Vitamin C (mg/100 gm)	Major elements (mg/100 gm)				Minor elements (mg/100 gm)			
		Na	K	Mg	Ca	Zn	Cu	Co	Mo
S. caseolaris health drink	63.11	392.42	406.24	138.47	38.12	1.86	1.27	BDL	BDL

Table 4.42 Health benefits of vitamin C

Human health benefit	Mechanism
(a) Allergy and asthma relief	Bronchial constriction is reduced
(b) Cancer prevention	Presence of antioxidant reduces the risk of cancer
(c) Cataract prevention	Replenishment of vitamin C in the lens and vitreous humor is a viable strategy for minimizing oxidative stress and reduce the risk of cataract formation
(d) Healing of wound	Vitamin C promotes faster healing of wounds and injuries because of its role in collagen production
(e) Diabetes control	Vitamin C supplementation may assist diabetics in controlling blood sugar levels and improving metabolism
(f) Immune system booster	Vitamin C increases white blood cell production and is important to immune system balance
(g) Neurotransmitter and hormone building	Vitamin C is critical to the conversion of certain substances into neurotransmitters, brain chemicals that facilitate the transmission of nerve impulses across a synapse (the space between neurons, or nerve cells). Such neurotransmitters as serotonin, dopamine, and nor epinephrine are responsible for the proper functioning of the central nervous system, and a deficiency of neurotransmitters can result in psychiatric illness. Vitamin C also helps the body manufacture adrenal hormones

In this analysis, Combined Alternative Livelihood Preference (CALP) was enumerated based on Alternative Livelihood Preference Score (ALPS) for various edible products of *S. caseolaris* fruit, computed as per the expression:

$$\text{CALP} = \text{ALPS1} + \text{ALPS2} + \text{ALPS3} + \text{ALPS4}$$

where, ALPS = Preference Rank (PR) × % of Vote.

The respondents provided the weightage to edibility of the *S. caseolaris* fruit (based on the value of CALP) as per the order Raw pulp of *S. caseolaris* fruit (530.0) > Health drink from *S. caseolaris* (415.7) > *S. caseolaris*-based jelly (272.6) > *S. caseolaris*-based curry (208.4) > *S. caseolaris*-based cookies (118.0). We observed considerable inclination of the local inhabitants towards *S. caseolaris*-based health drink, and hence a benefit cost analysis was crried out to scale-up this mangrove based emission free livelihood. However, nurseries of the species need to be developed for back up support of the raw materials. These nurseries will also act as sink of carbon dioxide during their growth phase.

Tables 4.44, 4.45, 4.46 and 4.47 present the Benefit Cost Analysis of *S. caseolaris*-based health drink in the low saline zone of Indian Sundarbans.

The supply chain of health drink is constantly undergoing a change. Industry regulations and quality control are the key issues for managing their supply chains and getting products from point A to point B while simultaneously maintaining compliance with continually evolving standards. The distribution of *S. caseolaris*–based health drink from the production center (*S. caseolaris* stands in the low saline

Table 4.43 *S. caseolaris* edibility preference of respondents in Indian Sundarbans based on Alternative Livelihood Preference Score (ALPS)

S. caseolaris fruit-based alternative livelihood	Policy maker (Respondent Type 1)		
	PR	% of Vote	ALPS1
Raw pulp of *S. caseolaris* fruit	5	26.8	134
S. caseolaris-based jelly	3	19.7	59.1
S. caseolaris-based curry	2	15.4	30.8
Health drink from *S. caseolaris*	4	24.4	97.6
S. caseolaris-based cookies	2	13.7	27.4
S. caseolaris fruit-based alternative livelihood	Researcher (Respondent Type 2)		
	PR	% of Vote	ALPS2
Raw pulp of *S. caseolaris* fruit	5	25.3	126.5
S. caseolaris-based jelly	4	20.3	81.2
S. caseolaris-based curry	3	16.2	48.6
Health drink from *S. caseolaris*	4	24.2	96.8
S. caseolaris-based cookies	3	14	42
S. caseolaris fruit-based alternative livelihood	Agriculturist (Respondent Type 3)		
	PR	% of Vote	ALPS3
Raw pulp of *S. caseolaris* fruit	5	27.3	136.5
S. caseolaris-based jelly	3	16.9	50.7
S. caseolaris-based curry	4	20.1	80.4
Health drink from *S. caseolaris*	4	25.2	100.8
S. caseolaris-based cookies	1	10.5	10.5
S. caseolaris fruit-based alternative livelihood	Local entreprenuer (Respondent Type 4)		
	PR	% of Vote	ALPS4
Raw pulp of *S. caseolaris* fruit	5	26.6	133
S. caseolaris-based jelly	4	20.4	81.6
S. caseolaris-based curry	3	16.2	48.6
Health drink from *S. caseolaris*	5	24.1	120.5
S. caseolaris-based cookies	3	12.7	38.1

Table 4.44 Fixed cost of health drink preparation

Sl. No.	Particulars	Amount (Rs.)
1	Room Rent @ Rs. 1,500 per month	18,000/-
2	Machinery and equipment	15,000/-
Total fixed cost		33,000/-

Table 4.45 Operational cost of health drink preparation

Sl. No.	Particulars	Amount (Rs.)
1	Procurement of *S. caseolaris* fruit (avg. wt. 20 gm)—400 kg @ Rs. 60 per kg	24,000/-
2	Electricity	10,000/-
3	Transportation	15,000/-
4	Sugar	3,000/-
5	Preservatives	255/-
6	Packing cost	5,000/-
7	Labour charges–2 heads @ Rs. 3,000 per head per month for 4 months	24,000/-
8	Miscellaneous	5,000/-
Total operational cost		86,255/-

Table 4.46 Total cost of health drink preparation

Sl. No.	Particulars	Amount (Rs.)
1	Fixed cost	33,000/-
2	Operational cost	86,255/-
3	Depreciation on fixed cost @10%/year	3,300/-
4	Interest on fixed investment @15%/year	4,950/-
Total cost		1,27,505/-

Table 4.47 Selling price of *S. caseolaris*-based Health drink

Sl. No.	Particulars	Amount (Rs.)
1	Health drink—556 bottle (each 1 lit) @ Rs. 280 per lit	1,55,680/-
Net income (Total sale – Total cost)		28,175/-

belt of Indian Sundarbans) to end consumers involves processors (for processing both raw fruits and health drink), traders, wholesalers, and consumers who purchase them from the shopping malls, medicinal shops or any other ordinary shops (Fig. 4.47).

4.3.2 *Biofertilizer Preparation from* **Azolla** *sp.*

In Indian Sundarban, the low saline pockets are restricted in the upstream region of the Hooghly estuary like Kakdwip (21° 52′ 26.50″ N and 88° 08′ 04.48″ E). The average salinity of the region is around 3.5 psu, which rises to 9.0 psu during the premonsoon season when there is no rainfall and the evaporation rate is quite high. During the monsoon season that starts from mid-June and extends up to October, the salinity drops to almost zero. The inland water bodies which are mostly freshwater

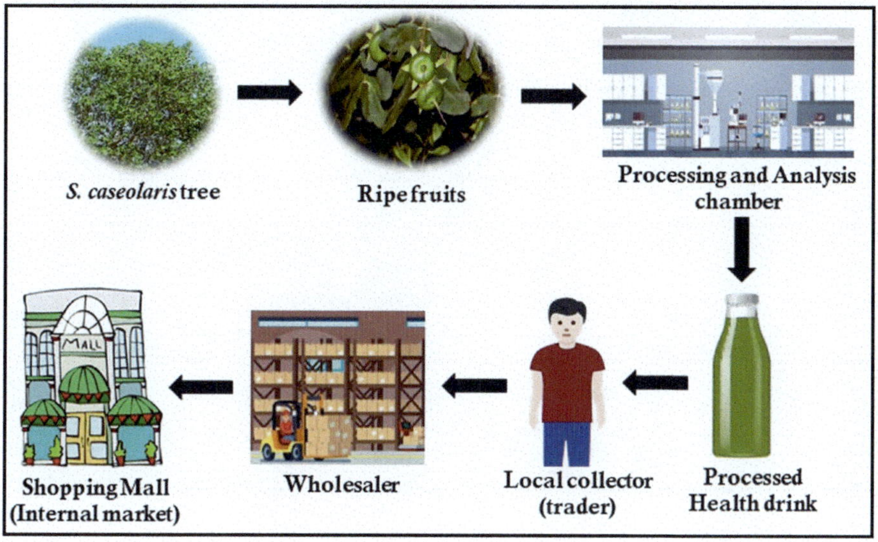

Fig. 4.47 Diagrammatic representation of supply chain in the *S. caseolaris*-based health drink secto

in nature act as ideal site for *Azolla* farming. The rain water harvested canals also provide suitable niche for *Azolla* propagation.

Azolla is an aquatic fern, which floats on the surface of water with their roots submerged. The species is unique because of its fast growth and can double the area it covers within 7–10 days. The species is also known for its unique carbon storage potential both in the early phase of the life cycle in congenial environment (45.7%) and under stressful condition (mostly due to high exposure to sunlight and high temperature), when it becomes reddish-brown in color (41.3%). Figure 4.48 highlights the carbon content in *Azolla* sp. available in the low saline belt of Indian Sundarbans in two different environmental conditions.

We analyzed the chemical composition of the species in three seasons (premonsoon, monsoon and postmonsoon) from a fresh water pond in Kakdwip during February 2022 and observed that the species has considerable nutritional value (Table 4.48).

Based on our observation, we initiated *Azolla* culture in the freshwater pond located in Kakdwip during March 2022. The culture has several advantages as (i) the species can grow easily in the wild, (ii) fast growing in nature, (iii) can store atmospheric nitrogen and carbon dioxide, (iv) efficient in solubilising elements like zinc, iron, manganese etc. and make these available to crop plants, (v) can be used to make biofertilizers, fish feed, cattle feed etc. (vi) needs less investment for farming.

We maintained few environmental criteria for *Azolla* farming as stated here.

Fig. 4.48 *Azolla* culture in the low saline belt of Indian Sundarbans

Table 4.48 Seasonal variation of chemical composition of *Azolla*

Substances (in %)	Premonsoon	Monsoon	Postmonsoon
Nitrogen	5.25	6.08	5.83
Phosphorus	0.52	0.71	0.59
Potassium	3.95	5.16	4.87
Calcium	0.85	1.21	0.93
Magnesium	0.60	0.72	0.64
Manganese	0.14	0.18	0.17
Iron	0.29	0.33	ND
Crude Fat	2.90	3.35	3.96
Sugar	3.39	3.31	3.42

1. The pond depth was ~60 cm, although for *Azolla* farming various depths were suggested by several researchers e.g. <5 cm (Sadeghi and Zarkami 2013), 5–10 cm (Hove 1989), 30–120 cm (Sadeghi et al. 2012), and 50 cm (Sadeghi et al. 2012).
2. The pH of our pond ranged between 5.5 and 6.8, although *Azolla* can adapt in different pH ranges like 4.5–7 (Wagner 1997), 5.3–5.8 (Katole et al. 2017), 4–10 (Hove 1989), 7.63 ± 0.31 (Sadeghi et al. 2012).
3. A photoperiod of 12 h was maintained, but reports of 20 h (Wagner 1997), and 5 h (Sadeghi et al. 2012) are available.
4. The temperature of the pond was around 35 °C during the day time.
5. Phosphorus was applied by using super phosphate and occasionally cow dung is used to maintain the nitrogen level.

Azolla sp. is a good fish and cattle feed that has the potential to serve as a source of income for the local inhabitants, but considering the nitrogen level of the species, we developed biofertilizer from the harvested *Azolla* through the following steps (Fig. 4.49).

Step 1 In this step, a pit of dimension 5 m × 2 m × 3 m was excavated.

Step 2 (A zone): The bottom most layer of the pit was filled with poultry excreta about 50 cm thick.

Step 3 (B zone): The 2nd layer from the bottom was filled with harvested *Azolla*. The thickness was 100 cm.

Step 4 (C zone): The 3rd layer from the bottom was covered/filled with cow dung of 50 cm.

Step 5 (D zone): The 4th layer above the cow dung was filled with *Azolla* and thickness of the layer was 50 cm.

Step 6 (E zone): The 5th layer was filled up with molasses and microbial inoculums (BIOJUICE) whose thickness was 25 cm.

We obtained an average yield of *Azolla*-based biofertilizer of 1.5 kg/day from our designed experiment.

Tables 4.49, 4.50, 4.51 and 4.52 present the Benefit Cost Analysis of *Azolla*-based Biofertilizer in the low saline zone of Indian Sundarbans.

The business model of *Azolla*-based biofertilizer simultaneously generate economic, social, and ecological value(s). These three values are related to the three P's i.e., People, Profit and Planet. Sustainability is embedded in these values. It is not just about earning money, but also caring for environment. *Azolla*-based biofertilizer, being a rich source of nitrogen has several merits like (i) improvement of soil fertility, (ii) reduction of weeds, (iii) increase of soil organic carbon, (iv) improvement of soil microbial load etc. As it is a substitute of urea, therefore the chemical contamination of environment can be avoided through application of *Azolla*-based biofertilizer. It has been established that each hectare of soil requires 20 tonnes of *Azolla* in dry conditions to maintain fertility. The supply chain of *Azolla*-based biofertilizer needs the farming of the species at the base of the chain (production volume and quality). In the next stage the harvested biomass is subject to microbial processing to form biofertilizer, which then reaches the market through traders and wholesalers (Fig. 4.50).

4.4 Take Home Messages

[A] Approximately 3 billion people, which is about half of the world's population live within 200 km of the world's coastline. Climate change has posed severe adverse impacts on this large coastal population by way of sea level rise, coastal erosion, inundations, storm floods, tidal water encroachment into estuaries and river systems, acidification of coastal waters, alteration of biodiversity,

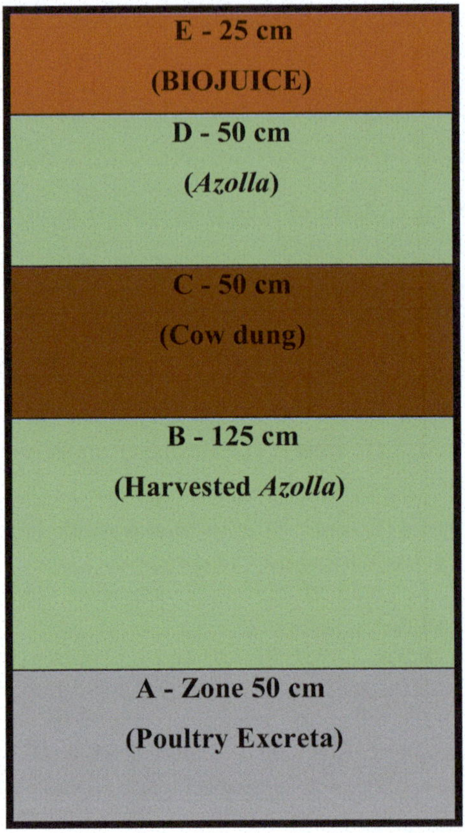

Fig. 4.49 Diagrammatic representation of biofertilizer development from *Azolla* sp.

Table 4.49 Fixed cost of *Azolla* farming

Sl. No.	Particulars	Amount (Rs.)
1	Trench cutting—2 Nos. @Rs. 1,000/trench	2,000/-
2	Silpauline tarpauline sheets—2 Nos. @Rs. 500/sheet	1,000/-
3	Polynet or shade @Rs. 1000/polynet	2,000/-
4	Wooden sticks/bamboo	2,000/-
Total fixed cost		7,000/-

contamination of freshwater reserves and food crops (with saline water), loss of properties etc. Latest figures exhibit that 90 per cent of the disasters are linked with climate change that incurs a loss of world economy by US$520 billion per year and has high probability to push about 26 million people into poverty or below the poverty line every year (https://www.un.org/sustainabledevelopment/wp-content/uploads/2017/05/Ocean-fact-sheet-package.pdf). The

Table 4.50 Operational cost of *Azolla*-based biofertilizer

Sl. No.	Particulars	Amount (Rs.)
1	Labour charge	1,600/-
2	Fertile soil—20 kg/trench @ Rs 100/kg	4,000/-
3	Cow dung—10 kg/trench—Rs. 15/kg	300/-
4	Super phosphate fertilizer—10 kg/trench @ Rs. 80/kg	1,600/-
5	Fresh *Azolla* culture—2 kg/trench @ Rs. 100/kg	400/-
6	Micronutrients	200/-
7	Miscellaneous	500/-
Total operational cost		8,600/-

Table 4.51 Total cost of *Azolla*-based biofertilizer

Sl. No.	Particulars	Amount (Rs.)
1	Fixed cost	7,000/-
2	Operational cost	8,600/-
Total cost		15,600/-

Table 4.52 Selling price of *Azolla*-based biofertilizer

Sl. No.	Particulars	Amount (Rs.)
1	*Azolla*-based biofertilizer—production volume = 90 kg from 60 days @ 1.5 kg/day; Selling price = Rs. 300/kg	27,000/-
Net income (Total sale – Total cost)		11,400/- (from 60 days)

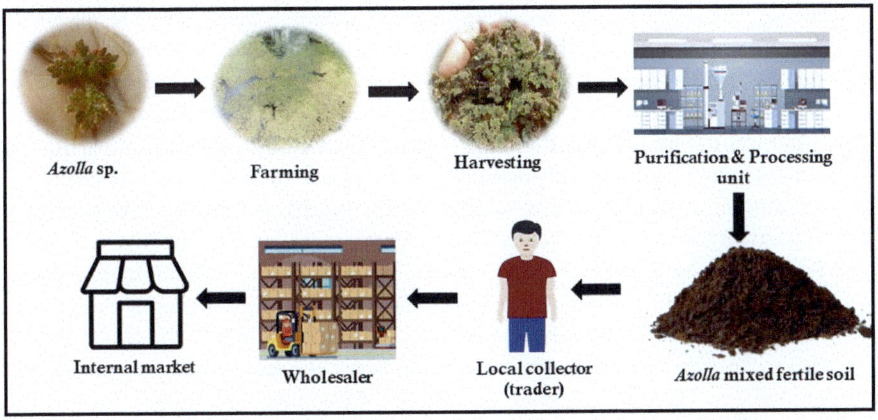

Fig. 4.50 Diagrammatic representation of supply chain in the *Azolla*-based biofertilizer sector

delta complex of Indian Sundarbans at the apex of Bay of Bengal is no exception to these dark facts. Several islands have shown massive erosion after the events of super—cyclones like Aila, Amphan, Yash etc.

[B] The concept of climate-resilient sustainable livelihoods implies not just what people do in order to live and survive comfortably for a short period of time, but the resources that provide them a long-term benefit for several generations. In addition to risk factor, different categories of sustainable livelihoods take in to account the management of resources coupled with alignment/non—alignment with the existing institutional support and policies related to livelihoods. In the domain of sustainable livelihood, resources are referred to as 'assets' or 'capitals' and are classified into several sub—categories like human capital (skills, education, and health), physical capital (produced investment goods), financial capital (money savings and access to loan), natural capital (land, water, flora, fauna, microbes etc.) and social capital (networks and association e.g., linkage of the entrepreneurs with the block and panchayet offices in case of Sundarbans).

[C] In order to develop a sustainable climate resilient livelihood model in the frame work of mangrove dominated Indian Sundarbans, a wide range of relevant data sets need to be explored considering the physico-chemical features of three major sectors (western, central, and eastern) of Indian Sundarbans like (i) Regional, sectoral, rural, and urban employment dynamics, (ii) Sector-wise ecological profile of Indian Sundarbans (preferably salinity-based spatial variation), (iii) Sector-wise existing livelihoods in Indian Sundarbans, (iv) Awareness on climate resilient mangrove—centric innovative livelihood, (v) Sector-wise Island business model prospects, and (vi) Existing institutional mechanism and policies related to livelihoods in the islands of Indian Sundarbans.

[D] Based on significant spatial variations of salinity (the major driver of biodiversity variation), there exists significant variation in alternative livelihood profile. The high saline central sector of Indian Sundarbans is favourable for halophyte farming and shrimp feed based on halophytes like *Suaeda* or *Salicornia*. These species grow luxuriantly in the supra-littoral zone of the intertidal mudflats in the high saline zone of Indian Sundarbans. However, in the medium saline zone, oyster culture and seaweed culture can be carried out, provided water quality management is scientifically done. The western and the eastern sectors of Indian Sundarbans are noted for low salinity, but in our discussion, we have primarily focused on the western sector as the eastern sector falls under the Reserve Forest (RF) area, where any type of manmade activities are strictly under control as per the Act of the Government Forest Department. Based on the availability of natural resources, *S. caseolaris*-based health drink and *Azolla*-based biofertilizer can serve as prospective livelihoods for the low saline western Indian Sundarbans. However, awareness of the local people coupled with institutional support is the major driver to drag the climate resilient alternative livelihoods of Indian Sundarbans from the hadal zone to the surface.

References

Adolf VI, Jacobsen SE, Shabala S (2013) Salt tolerance mechanisms in quinoa (*Chenopodium quinoa* Willd.). Environ Exp Bot 92:43–54

Agarwal S, Albeshr MF, Mahboobb S, Atique U, Pramanick P, Mitra A (2022) Bioaccumulation factor (BAF) of heavy metals in green seaweed to assess the phytoremediation potential. J King Saud Univ Sci (Elsevier) 34(5):102078(1–7). https://doi.org/10.1016/j.jksus.2022.102078

Ahmed AK, Johnson KA (2000) Horticultural development of Australian native edible plants. Aust J Bot 48:417–426

Amin G, Trivedi S, Pramanick P, Mitra A (2018) Spatio-temporal variations of Heavy metal levels in green seaweed inhabiting Indian Sundarbans. Sci Fed J Sep Tech 1(1):1–3

Bandaranayke WM (2002) Bioactivities, bioactive compounds and chemical constituents of mangrove plant. Wetlands Ecol Manag 10:421–452

Banerjee K, Chakraborty S, Paul R, Bal G, Zaman S, Pramanick P, Amin G, Fazli P, Mitra A (2015) Macrobenthic molluscan diversity in the major intertidal zones of the maritime states of India. J Environ Sci Pollut Res 1(1):8–11

Ben Hamed K, Debez A, Chibani F, Abdelly C (2004) Salt response of *Crithmum maritimum*, an oleagineous halophyte. Trop Ecol 45:151–159

BOSTID (1990) Saline agriculture salt-tolerant plants for developing countries. National Academy Press, Washington, DC. http://pdf.usaid.gov/pdfdocs/PNABH313.pdf

Carlsson R, Clarke EMW (1983) Atriplex hortensis L. as a leafy vegetable and as a leaf protein concentrate plant. Plant Foods Hum Nutr 33:127–133

Chaudhuri AB, Choudhury A (1994) Mangroves of the Sundarbans. Volume 1: India. World Conservation Union, Gland, 247

de Vos AC, Broekman R, Guerra CCD, van Rijsselberghe M, Rozema J (2013) Developing and testing new halophyte crops: a case study of salt tolerance of two species of the Brassicaceae, *Diplotaxis tenuifolia* and *Cochlearia officinalis*. Environ Exp Bot 92:154–164

Debez A, Saadaoui D, Slama I, Huchzermeyer B, Abdelly C (2010) Responses of Batis maritima plants challenged with up to two-fold seawater NaCl salinity. J Plant Nutr Soil Sci 173:291–299

Dubois M, Gilles KA, Hamilton JK, Rebers PA, Smith F (1956) Colorimetric methods for determination of sugars and related substances. Analyt Chem 28:350–356

Ellis F (1998) Survey article: household strategies and rural livelihood diversification. J Dev Stud 35(1):1–38

Ellis F (2000) Rural livelihoods and diversity in developing countries. Oxford University Press, Oxford

Essaidi I, Brahmi Z, Snoussi A, Ben Haj Koubaier H, Casabianca H, Abe N, El Omri A, Chaabouni MM, Bouzouita N (2013) Phytochemical investigation of Tunisian *Salicornia herbacea* L., antioxidant, antimicrobial and cytochrome P450 (CYPs) inhibitory activities of its methanol extract. Food Control 32:125–133. https://doi.org/10.1016/j.foodcont.2012.11.006

Felger RS, McRoy CP (1975) Seagrasses as potential food plants. In: Somem CF (ed) Seed bearing halophytes as food plants. University of Delaware, Newark, Delaware, US, College of Marine Studies, pp 62–69

Felger RS, Moser MB (1976) Seri Indian food plants: desert subsistence without agriculture. Ecol Food Nutr 5:13–27

Flowers TJ, Gracia A, Koyama M, Yeo AR (1997) Breeding for salt tolerance in crop plants. The role of molecular biology. Acta Physiol Plant 19:427–433

Folch J, Lees M, Solam-Stanley GH (1957) A simple method for the isolation and purification of claot lipid from animal tissue. J Biol Chem 226:497–509

Frank W (1982) Vitamin C in sea fennel (*Crithmum maritimum*), an edible wild plant. Econ Bot 36(2):163–165

Hove CV (1989) *Azolla*: and its multiple uses with emphasis on Africa. Food and Agriculture Organization of the United Nations, Rome, Italy

https://run.unl.pt/bitstream/10362/25208/1/Vieira_etall_2017.pdf

https://worldpopulationreview.com/countries
https://www.un.org/sustainabledevelopment/wp-content/uploads/2017/05/Ocean-fact-sheet-package.pdf
Katole SB, Lende SR, Patil SS (2017) A review on potential livestock feed: *Azolla*. Livest Res Int 05:01–09
Koziol MJ (1992) Chemical composition and nutritional evaluation of quinoa (*Chenopodium quinoa* Willd.). J Food Compost Anal 5:35–68
Lokhande VH, Nikam TD, Suprasanna P (2009) *Sesuvium portulacastrum* (L.) L. a promising halophyte: cultivation, utilization and distribution in India. Genet Resour Crop Evol 56:741–747. https://doi.org/10.1007/s10722-009-9435-1
Maat GJR (2004) Scurvy in adults and youngsters: the Dutch experience. A review of the history and pathology of a disregarded disease. Int J Osteoarchaeol 14:77–81
Mitra A (2013) Sensitivity of mangrove ecosystem to changing climate. Springer, New Delhi, Heidelberg, New York, Dordrecht, London
Mitra A (2023) Impact of COVID-19 lockdown on environmental health: exploring the situation of the lower Gangetic delta. Springer ISBN 978-3-031-27242-4 (eBook), vol XIII, pp 310. https://doi.org/10.1007/978-3-031-27242-4
Mitra A, Zaman S (2015) Blue carbon reservoir of the blue planet published by Springer, ISBN 978-81-322-2106-7. XII. https://doi.org/10.1007/978-81-322-2107-4
Mitra A, Zaman S, Pramanick P (2022) Blue economy in Indian Sundarbans: exploring livelihood opportunities. Springer ISBN 978-3-031-07908-5 (e-Book), vol XIV, pp 403. https://doi.org/10.1007/978-3-031-07908-5
Mukhopadhyay PK, Rangacharyulu PV, Mitra G, Jana BB (2003) Applied nutrition in freshwater prawn, *Macrobrachium rosenbergii*, culture. J Appl Aquac 13:317–340
Nambiar VS, Dhaduk JJ, Sareen N, Shahu T, Desai R (2011) Potential functional implications of pearl millet (*Pennisetum glaucum*) in health and disease. J Appl Pharma Sci 1(10):62–67
Pearlsteina SL, Felger RS, Glenn EP, Harringtond J, Al-Ghanem KA, Nelsona SG (2012) NyPa (*Distichlis palmeri*): a perennial grain crop for saltwater irrigation. J Arid Environ 82:60–70
Pornpitakdamrong A, Sudjaroen Y (2014) Seablite (*Suaeda maritima*) product for cooking, Samut Songkram province, Thailand. Food Nutri Sci 5:850–856
Pramanick P, Bera D, Banerjee K, Zaman S, Mitra A (2016) Seasonal variation of proximate composition of common seaweeds in Indian Sundarbans. Int J Life Sci Sci Res 2(5):570–578
Pramanick P, Zaman S, Rudra T, Guha A, Mitra A (2015) Heavy metals in a dominant seaweed species from the islands of Indian Sundarbans. Int J Life Sci Pharm Res 5(2):64–71
Risi JC, Galway NW (1984) The Chenopodium grains of the Andes: Inca crops for modern agriculture. Adv Appl Biol 10:145–216
Roy Chowdhury G, Pal N, Zaman S, Saha A, Mitra A (2017) Shrimp seed collection in Indian Sundarban Estuaries: a threat to overall estuaries ecosystem services. J Environ Social Sci 4(1):1–3
Sadeghi R, Zarkami R (2013) A review of some ecological factors affecting the growth of *Azolla* spp. Casp J Envirion Sci 11:65–76
Sadeghi R, Zarkami R, Sabetraftar K, Van Damme P (2012) Application of classification trees to model the distribution pattern of a new exotic species *Azolla filiculoides* (Lam.) at Selkeh Wildlife Refuge. Anzali Wetland. Iran Ecol Modell 243:8–17. https://doi.org/10.1016/j.ecolmodel.2012.06.011
Shannon MC, Grieve CM (1999) Tolerance of vegetable crops to salinity. Sci Hortic 78:5–38
Simopoulos AP (2004) Omega-3 fatty acids and antioxidants in edible wild plants. Bio Res 37:263–277
Singh D, Buhmann AK, Flowers TJ et al (2014) *Salicornia* as a crop plant in temperate regions: selection of genetically characterized ecotypes and optimization of their cultivation conditions. AoB Plants. https://doi.org/10.1093/aobpla/plu071

Słupski J, Achrem-Achremowicz J, Lisiewska Z, Korus A (2010) Effect of processing on the amino acid content of New Zealand spinach (*Tetragonia tetragonioides* Pall. Kuntze). Int J Food Sci Technol 45:1682–1688

Smillie C (2015) *Suaeda* spp. as a biomonitor of Cu and Zn in salt marsh sediments. Ecol Indic 56:70–78. https://doi.org/10.1016/j.ecolind.2015.03.010

Tanaka T (1976) Tanaka's cyclopedia of edible plants of the world. Keigaku publishing, Tokyo

Tecator (1987) Determination of Kjeldahl nitrogen content with Kjeltec system 1026 Appication Note 86/87. 1987.02.18. Kjeltec 1026 Manual. Tecator, Sweden

Ventura Y, Wuddineh WA, Myrzabayeva M, Alikulov Z, Khozin-Goldberg I, Shpigel M, Samocha TM, Sagi M (2011) Effect of seawater concentration on the productivity and nutritional value of annual *Salicornia* and perennial *Sarcocornia* halophytes as leafy vegetable crops. Sci Hortic 128:189–196

Wagner GM (1997) Azolla: a review of its biology and utilization. Bot Rev 63:1–26. https://doi.org/10.1007/BF02857915

Yamamoto K, Oguri S, Chiba S, Momonoki YS (2009) Molecular cloning of acetylcholinesterase gene from *Salicornia europaea* L. Plant Signal Behav 4:361–366

Yazici I, Turkan I, Sekmen AH, Demiral T (2007) Salinity tolerance of purslane (*Portulaca oleracea* L.) is achieved by enhanced antioxidative system, lower level of lipid peroxidation and proline accumulation. Environ Exp Bot 61:49–57

Yensen NP, Yensen SB, Weber CW (1985) A review of *Distichlis* spp. for production and nutritional value. In: Whitehead EE, Hutchenson EF, Timmermann BN, Varady RG (eds) Arid land today and tomorrow. Westview Press, Boulder, CO, pp 809–822

Zerai DB, Glenn EP, Chatervedi R, Lu Z, Mamood AN, Nelson SG, Ray DT (2010) Potential for the improvement of *Salicornia bigelovii* through selective breeding. Ecol Eng 36:730–739

Chapter 5
Blue Carbon Credit: An Approach Towards Net Zero Livelihood

Contents

5.1 Carbon Credit Through True Mangrove Species

Mangrove vegetations play an important role in mitigating climate change through carbon sequestration. These halophytes absorb and store carbon dioxide from the atmosphere in their biomass and in the soil/sediment beneath them. This makes the mangrove ecosystem a potential natural carbon sink and a unique reservoir to gain carbon credit.

Carbon credits are tradeable permits that represent a certain amount of carbon dioxide or other greenhouse gas emissions that have been prevented, reduced, or removed from the atmosphere. Companies or individuals have the right to purchase these credits in order to offset their own carbon emissions, contributing to the global effort to mitigate climate change.

Mangrove conservation and restoration projects can generate carbon credits through quantification of the amount of carbon stored in the mangrove ecosystem. This involves measuring the carbon content of the mangrove trees, the sediment, and the associated fauna and flora, and using established methodologies to calculate the amount of carbon dioxide equivalent that has been sequestered (*Vide* Annexure 5.1

A. Mitra et al., *Climate Resilient Innovative Livelihoods in Indian Sundarban Delta*,
https://doi.org/10.1007/978-3-031-42633-9_5

for the standard methodologies we have used). These carbon credits can then be sold to companies or individuals who want to offset their carbon emissions. The revenue generated from the sale of these credits can be used to finance further mangrove conservation and restoration efforts by the local communities, creating a positive feedback loop that benefits both the environment and the local communities, preferably the coastal population or the people living in the delta complex and estuarine region.

Overall, carbon credits through mangrove plants offer a promising avenue for mitigating climate change while promoting sustainable development in coastal regions.

Mangroves are typically found along tropical and subtropical coastlines between about 25° N and 25° S. The common halophytes thriving within the mangrove habitats of the world encompasses 31 families under which 48 genera and 96 species are included as documented in the site http://www.bsienvis.nic.in/Database/IndianMangroves_3941.aspx#WORLDDISTRIBUTIONOFMANGROVES (Table 5.1).

In India mangrove forests in are found along the coastline of 9 States and 4 Union Territories (Fig. 5.1).

Mangroves in the Indian sub-continent have been extensively studied from the taxonomic point of view. However, slight discrepancies exist in relation to the number of true mangrove species (Table 5.2). For instance, Singh and Garge (1993) listed 32 mangrove species; Dagar et al. (1993) listed 36 mangrove species, while Naskar (2004); Selvam et al. (2002) and Kathiresan and Rajendran (2005) listed 43, 35 and 39 species respectively. In the recent past Mandal and Naskar (2008) and Sanjappa et al. (2011) attempted to review the mangrove flora of India, but their studies do not give a complete picture of the true mangrove species of India. Mandal and Naskar (2008) recognized a total of 82 mangrove species belonging to 52 genera and 36 families reported by various studies and tried to classify them into major mangrove and mangrove associates based on modified morphological and anatomical characters of leaves, stems, roots, and reproductive organs. They listed 30 major mangrove species (true mangrove species) in India. The shortfalls of their results are species such as *Cynometra iripa, Excoecaria indica, Pemphis acidula* and *Acrostichum speciosum* were not included in their list and *Acanthus volubilis, Brownlowia tersa, Acrostichum aureum* and *Dolichandrone spathacea* were classified as mangrove associates. Similarly Anonymous (2008) and Kathiresan (2008) recognized 39 true mangrove species with the exclusion of *Acanthus volubilis, Aglaia cucullata* and *Phoenix paludosa* and they included *Cynometra ramiflora*, a species not considered as true mangrove species globally. Sanjappa et al. (2011) reported 68 true mangrove species, in their list of coastal plants and salt marsh species and *Caesalpinia crista, Cerbera manghas, Clerodendrum inerme, Dalbergia spinosa, Thespesia populneoides* etc. were included as true mangroves.

The impact in terms of confusion of including mangrove associates or beach vegetation and species unlikely to be present when reporting the total number of mangrove species in a given habitat has been discussed in detail by Jayatissa et al. (2002). Hence, in this chapter we have separated the entire halophytes of Indian

Table 5.1 World distributions of common mangrove and associate species

Sl. No.	Species	
1	*Acanthus ebracteatus*	
2	*Acanthus ilicifolius*	
3	*Acanthus volubilis*	
4	*Cerbera manghas*	
5	*Nypa fruticans*	

(continued)

Table 5.1 (continued)

Sl. No.	Species	
6	*Phoenix paludosa*	
7	*Finlaysonia obovata*	
8	*Sarcolobus globosus*	
9	*Sarcolobus carinatus*	
10	*Tylophora tenuis*	

(continued)

Table 5.1 (continued)

Sl. No.	Species	
11	*Avicennia alba*	
12	*Avicennia bicolour*	
13	*Avicennia germinans*	
14	*Avicennia integra*	
15	*Avicennia marina*	

(continued)

Table 5.1 (continued)

Sl. No.	Species	
16	*Avicennia var. australasica*	
17	*Avicennia var. eucalyptifolia*	
18	*Avicennia officinalis*	
19	*Avicennia rumphiana*	
20	*Avicennia schaueriana*	

(continued)

Table 5.1 (continued)

Sl. No.	Species	
21	*Dolichandrone spathacea*	
22	*Camptostemon philippinensis*	
23	*Camptostemon schultzii*	
24	*Caesalpinia bonduc*	
25	*Caesalpinia cristata*	

(continued)

Table 5.1 (continued)

Sl. No.	Species	
26	*Cynometra iripa*	
27	*Cynometra ramiflora*	
28	*Intsia bijuga*	
29	*Conocarpus erectus*	
30	*Laguncularia racemosa*	

(continued)

Table 5.1 (continued)

Sl. No.	Species	
31	*Lumnitzera littorea*	
32	*Lumnitzera racemosa*	
33	*Ipomoea tuba*	
34	*Fimbristylis ferruginea*	
35	*Scirpus littoralis*	

(continued)

Table 5.1 (continued)

Sl. No.	Species	
36	*Excoecaria agallocha*	
37	*Excoecaria indica*	
38	*Excoecaria ovate*	
39	*Flagellaria indica*	
40	*Pemphis acidula*	

(continued)

Table 5.1 (continued)

Sl. No.	Species	
41	*Hibiscus tiliaceous*	
42	*Thespesia populneoide*	
43	*Thespesia populnea*	
44	*Aglaia cucullata*	
45	*Xylocarpus granatum*	

(continued)

Table 5.1 (continued)

Sl. No.	Species	
46	*Xylocarpus mekongensis*	
47	*Xylocarpus moluccensis*	
48	*Aegiceras corniculatum*	
49	*Aegiceras floridum*	
50	*Osbornia octodonta*	

(continued)

Table 5.1 (continued)

Sl. No.	Species	
51	*Dalbergia spinosa*	
52	*Derris scandens*	
53	*Derris heterophylla*	
54	*Mucuna gigantea*	
55	*Pelliciera rhizophorae*	

(continued)

Table 5.1 (continued)

Sl. No.	Species	
56	*Aegialitis rotundifolia*	
57	*Aegialitis annulata*	
58	*Myriostachya wightiana*	
59	*Porteresia coarctata*	
60	*Urochondra setulosa*	

(continued)

Table 5.1 (continued)

Sl. No.	Species	
61	*Acrostichum aureum*	
62	*Acrostichum danaeifolium*	
63	*Acrostichum speciosum*	
64	*Bruguiera cylindrica*	
65	*Bruguiera exaristata*	

(continued)

Table 5.1 (continued)

Sl. No.	Species	
66	*Bruguiera gymnorhiza*	
67	*Bruguiera hainesii*	
68	*Bruguiera parviflora*	
69	*Bruguiera sexangula*	
70	*Ceriops australis*	

(continued)

Table 5.1 (continued)

Sl. No.	Species	
71	*Ceriops decandra*	
72	*Ceriops tagal*	
73	*Kandelia candel*	
74	*Rhizophora apiculata*	
75	*Rhizophora harrisonii*	

(continued)

Table 5.1 (continued)

Sl. No.	Species	
76	*Rhizophora mangle*	
77	*Rhizophora mucronata*	
78	*Rhizophora racemosa*	
79	*Rhizophora samoensis*	
80	*Rhizophora stylosa*	

(continued)

Table 5.1 (continued)

Sl. No.	Species	
81	*Scyphiphora hydrophyllacea*	
82	*Merope angulata*	
83	*Sonneratia alba*	
84	*Sonneratia apetala*	
85	*Sonneratia caseolaris*	

(continued)

Table 5.1 (continued)

Sl. No.	Species	
86	*Sonneratia griffithii*	
87	*Sonneratia lanceolata*	
88	*Sonneratia ovate*	
89	*Sonneratia gulngai*	
90	*Sonneratia urama*	

(continued)

Table 5.1 (continued)

Sl. No.	Species	
91	*Heritiera fomes*	
92	*Heritiera globosa*	
93	*Heritiera kanikensis*	
94	*Heritiera littoralis*	

(continued)

Table 5.1 (continued)

Sl. No.	Species	
95	*Brownlowia tersa*	
96	*Clerodendrum inerme*	

Source Mitra et al. (2022)

Sundarbans in to two separate units, namely true mangrove species and mangrove associate species.

The delta complex of Indian Sundarbans sustains 34 species of true mangroves (Table 5.3).

True mangrove flora are known for their natural potential to absorb carbon dioxide from the ambient media and store them in their vegetative and reproductive parts that finally can be aligned with carbon credit. There is an estimated 13.76 Mha of mangrove forests world-wide (Bunting et al. 2018) with some alterations in figures (Mitra 2013). According to Zeng et al. (2021) "approximately 20% (2.6–2.7 Mha) of 13.76 Mha is potentially investible for carbon finance projects, based on the probability of imminent threat, and approximately 10% will be financially sustainable-contributing up to 29.8 $MtCO_2$-e $year^{-1}$".

IPCC (2014) forward the view that "the total amount of investible mangrove blue carbon would contribute to climate change mitigation at a rate of 33.8 ± 5.1 $MtCO_2$-e $year^{-1}$, which is equivalent to approximately 0.13% of annual global carbon dioxide emissions, or 1.02% of annual average carbon dioxide emissions from Forestry and Other Land Use (FOLU) between 2002 and 2011". Based on the country-wise extent of mangroves, their biomass and carbon mitigation potential, and Net Present Value (NPV) in US$ $year^{-1}$, Southeast Asia, collectively accounts for 60% of global volumes (Table 5.4). In this context, the three major countries in Southeast Asia are Indonesia (11.3 ± 1.9 $MtCO_2$-e $year^{-1}$), Myanmar (2.9 ± 0.4 $MtCO_2$-e $year^{-1}$), and Vietnam (1.6 ± 0.3 $MtCO_2$-e $year^{-1}$).

Fig. 5.1 Distribution of mangrove forest in India

The calculations carried out to estimate the financial viability of mangrove blue carbon in Table 5.4 are based on the following assumptions:

(1) Uniform mangrove-based carbon price of US$ 5 t^{-1} CO_2-e for the first 5 years.
(2) 5% annual price appreciation for a 30 years project time fame.
(3) Risk-adjusted discount rate 10% for estimating Net Present Values (NPVs) for the Return of Investment (ROI) of mangrove blue carbon projects.

The cost per tonne of mangrove CO_2-e (carbon dioxide equivalent) can vary depending on the specific project, location, and market conditions. Mangrove conservation and restoration projects are often implemented to sequester carbon dioxide and other greenhouse gases, and the cost per tonne of CO_2-e can be influenced by various factors such as the type and scale of the project, the cost of implementing the project, and the value of carbon credits in the relevant carbon markets.

Table 5.2 List of true mangrove species reported by various authors in India

Species names	Singh and Garge (1993)	Dagar et al. (1993)	Naska (2004)	Selvam et al. (2002)	Kathiresan and Rajendran (2005)	Mandal and Naskar (2008)	Kathiresan (2008)
Acanthus ebracteatus	*	*	*		*		*
A. ilicifolius	*	*	*	*	*	*	*
A. volubilis	*	*	*				
Acrostichum aureum				*	*		*
A. speciosum					*		*
Aegialitis rotundifolia	*	*	*	*	*	*	*
Aegiceras corniculatum	*	*	*	*	*	*	*
Aglaia cucullata			*	*	*		
Avicennia alba	*	*	*	*	*	*	*
A. marina	*	*	*	*	*	*	*
A. marina var. acutissima			*				
A. officinalis	*	*	*	*	*	*	*
Atalantia correae			*				
Brownlowia tersa			*				
Bruguiera cylindrica	*	*	*	*	*	*	*
B. gymnorrhiza	*	*	*	*	*	*	*
B. parviflora	*	*	*	*	*	*	*
B. sexangula	*	*	*	*	*	*	*
Ceriops decandra	*	*	*	*	*	*	*
C. tagal	*	*	*	*	*	*	*
Cynometra iripa		*	*	*			
Cy. Ramiflora		*	*				*
Dalbergia spinosa			*				
Dolichandrone spathacea					*		*
Excoecaria agallocha	*	*	*	*	*	*	*
E. indica					*		*

(continued)

Table 5.2 (continued)

Species names	Singh and Garge (1993)	Dagar et al. (1993)	Naska (2004)	Selvam et al. (2002)	Kathiresan and Rajendran (2005)	Mandal and Naskar (2008)	Kathiresan (2008)
Heritiera fomes	*	*	*	*	*	*	*
H. littoralis	*	*	*	*	*	*	*
H. kanikensis				*	*		
Kandelia candel	*	*	*	*	*	*	*
Lumnitzera littorea	*	*	*		*	*	*
L. racemosa	*	*	*	*	*	*	*
Nypa fruticans	*	*	*	*	*	*	*
Pemphis acidula				*	*		*
Phoenix paludosa	*	*	*			*	
Rhizophora × *annamalayana*					*		*
R. apiculata	*	*	*	*	*	*	*
R. mucronata	*	*	*	*	*	*	*
R. × *lamarckii*	*	*	*	*	*		*
R. stylosa	*	*	*	*	*	*	*
Sarcolobus carinatus			*				
S. globosus			*				
Scyphiphora hydrophyllacea	*	*	*	*	*	*	*
Sonneratia alba	*	*	*	*	*	*	*
S. apetala	*	*	*	*	*	*	*
S. caseolaris		*	*	*	*	*	*
S. griffithii		*	*	*	*	*	*
Xylocarpus granatum	*	*	*	*	*	*	*
X. moluccensis	*	*	*	*	*		
X. mekongensis	*	*	*	*		*	*
Total number of species reported	32	36	43	35	39	30	39

Table 5.3 True mangrove floral species in Indian Sundarbans

Scientific Name	Local Name	Family
Acanthus ilicifolius	Hargoja	Acanthaceae
Acanthus volubilis	Lata hargoja	Acanthaceae
Aegiceras corniculatum	Khalsi	Myrsinaceae
Aegialites rotundifolia	Tora	Plumbaginaceae
Amoora cucullata	Amur	Meliaceae
Avicennia alba	Peara bain	Avicenniaceae
Avicennia marina	Kalo bain	Avicenniaceae
Avicennia officinalis	Sada bain	Avicenniaceae
Brownlowia tersa	Lata	Tiliaceae
Bruguiera cylindrica	Bakul	Rhizophoraceae
Bruguiera gymnorhiza	Kankra	Rhizophoraceae
Bruguiera hexangula	Kankra	Rhizophoraceae
Bruguiera parviflora	Kankra bakul	Rhizophoraceae
Ceriops decandra	Goran	Rhizophoraceae
Ceriops tagal	Math goran	Rhizophoraceae
Cynometra ramiflora	Singara	Caesalpiniaceae
Derris trifoliata	Kalilata	Fabaceae
Derris umbellatum	–	Fabaceae
Excoecaria agallocha	Genwa	Euphorbiaceae
Excoecaria bicolor	–	Euphorbiaceae
Heritiera fomes	Sundari	Sterculiaceae
Hibiscus tortuosus	Paras	Malvaceae
Kandelia candel	Garia	Rhizophoraceae
Lumnitzera racemosa	Kirpa	Combretaceae
Nypa fruticans	Golpata	Palmae or Arecaceae
Phoenix paludosa	Hental	Palmae or Arecaceae
Rhizophora apiculata	Garjan	Rhizophoraceae
Rhizophora mucronata	Garjan	Rhizophoraceae
Sonneratia apetala	Keora	Sonneratiaceae
Sonneratia caseolaris	Ora	Sonneratiaceae
Tamarix dioica	Nona jhau	Tamaricaceae
Tamarix gallica	Nona jhau	Tamaricaceae
Xylocarpus granatum	Dhundul	Meliaceae
Xylocarpus mekongensis	Passur	Meliaceae

Table 5.4 Global, regional, and country-level estimates of the extent, climate mitigation potential, and return on investment (based on net present value) of financially viable mangrove blue carbon

Locations	Extent (ha)	Climate mitigation potential (tCO_2-e $year^{-1}$)	Net present value (US$ $year^{-1}$)
World	1,054,900 (±78,000)	26,164,000 (±4,664,000)	1,188,889,000 (±241,216,000)
America	58,800 (±17,800)	1,871,000 (±718,000)	47,749,000 (±21,718,000)
Brazil	17,300 (±4,900)	678,000 (±217,000)	15,645,000 (±6,071,000)
Cuba	10,000 (±5,400)	332,000 (±224,000)	6,704,000 (±6,548,000)
Colombia	5700 (±1,500)	196,000 (±60,000)	5,416,000 (±1,993,000)
French Guyana	4800 (±400)	96,000 (±16,000)	3,932,000 (±813,000)
Guyana	1300 (±100)	96,000 (±16,000)	3,932,000 (±815,000)
Venezuela	3000 (±700)	131,000 (±37,000)	2,962,000 (±1,037,000)
Nicaragua	3900 (±600)	51,000 (±14,000)	1,760,000 (±645,000)
Suriname	2800 (±1,200)	61,000 (±31,000)	1,369,000 (±905,000)
Ecuador	2100 (±600)	51,000 (±18,000)	1,322,000 (±576,000)
Honduras	2600 (±500)	38,000 (±11,000)	1,286,000 (±452,000)
Haiti	1900 (±400)	23,000 (±10,000)	1,056,000 (±537,000)
Panama	1000 (±300)	45,000 (±15,000)	796,000 (±346,000)
El Salvador	1000 (±300)	16,000 (±5,000)	561,000 (±167,000)
Costa Rica	200 (±100)	10,000 (±5,000)	211,000 (±106,000)
Jamaica	200 (±100)	10,000 (±5,000)	196,000 (±154,000)
Mexico	400 (±400)	14,000 (±16,000)	188,000 (±242,000)
Belize	300 (±100)	7000 (±3,000)	128,000 (±86,000)
Cayman Islands	200 (±100)	6000 (±6,000)	106,000 (±138,000)

(continued)

Table 5.4 (continued)

Locations	Extent (ha)	Climate mitigation potential (tCO_2-e $year^{-1}$)	Net present value (US$ $year^{-1}$)
Dominican Republic	200 (±100)	6000 (±4,000)	100,000 (±114,000)
Guatemala	100 (±0)	2000 (±0)	48,000 (±17,000)
Africa	240,500 (±14,900)	3,940,000 (±597,000)	211,323,000 (±35,007,000)
Madagascar	71,600 (±1,600)	1,557,000 (±158,000)	92,629,000 (±9,997,000)
Nigeria	73,100 (±8,300)	1,056,000 (±250,000)	48,202,000 (±13,821,000)
Cameroon	33,500 (±1,100)	524,000 (±76,000)	29,597,000 (±4,707,000)
Tanzania	13,500 (±400)	391,000 (±38,000)	22,035,000 (±2,348,000)
Guinea Bissau	15,800 (±1,200)	107,000 (±19,000)	4,867,000 (±1,089,000)
Mozambique	7500 (±300)	63,000 (±9,000)	3,326,000 (±540,000)
Sierra Leone	8900 (±500)	60,000 (±12,000)	3,153,000 (±711,000)
Gabon	1400 (±300)	52,000 (±13,000)	1,611,000 (±427,000)
Angola	900 (±100)	29,000 (±4,000)	1,294,000 (±214,000)
Kenya	2000 (±300)	29,000 (±6,000)	1,239,000 (±316,000)
Equatorial Guinea	500 (±0)	19,000 (±2,000)	1,093,000 (±116,000)
Guinea	9300 (±300)	19,000 (±2,000)	1,093,000 (±117,000)
Ghana	800 (±100)	16,000 (±4,000)	580,000 (±182,000)
Gambia	900 (±100)	5000 (±1,000)	181,000 (±67,000)
Senegal	500 (±200)	4000 (±2,000)	107,000 (±64,000)
Liberia	100 (±0)	2000 (±0)	96,000 (±14,000)
Somalia	100 (±0)	1000 (±0)	33,000 (±11,000)
Asia	703,600 (±42,600)	19,153,000 (±3,183,000)	874,468,000 (±175,033,000)
Indonesia	325,400 (±21,100)	11,262,000 (±1,889,000)	513,327,000 (±106,538,000)

(continued)

Table 5.4 (continued)

Locations	Extent (ha)	Climate mitigation potential (tCO_2-e $year^{-1}$)	Net present value (US$ $year^{-1}$)
Myanmar	124,500 (±3,100)	2,946,000 (±404,000)	157,135,000 (±24,841,000)
Vietnam	75,900 (±3,200)	1,635,000 (±272,000)	75,208,000 (±16,171,000)
Bangladesh	98,900 (±2,800)	808,000 (±75,000)	42,079,000 (±4,484,000)
Malaysia	23,500 (±5,000)	940,000 (±256,000)	24,854,000 (±9,019,000)
Andaman and Nicobar	6100 (±100)	444,000 (±47,000)	24,284,000 (±2,966,000)
Philippines	12,000 (±1,500)	511,000 (±105,000)	17,600,000 (±4,836,000)
Thailand	8800 (±1,700)	262,000 (±66,000)	6,623,000 (±2,312,000)
India	18,900 (±2,800)	192,000 (±43,000)	6,553,000 (±1,918,000)
Cambodia	7200 (±300)	116,000 (±16,000)	5,722,000 (±991,000)
Pakistan	2300 (±700)	24,000 (±10,000)	715,000 (±394,000)
Sri Lanka	200 (±100)	10,000 (±4,000)	239,000 (±139,000)
Oceania	51,900 (±2,700)	1,199,000 (±156,000)	55,349,000 (±8,833,000)
Papua New Guinea	48,600 (±2,500)	1,009,000 (±134,000)	47,240,000 (±7,706,000)
Solomon Islands	3200 (±100)	189,000 (±22,000)	8,091,000 (±1,235,000)

Source Zeng et al. (2021)

The cost per tonne of CO_2-e for mangrove conservation and restoration projects can range from a few dollars to several tens of dollars. However, it is important to note that the market for carbon credits is extremely dynamic and subject to various economic and regulatory factors, which can impact the value of carbon credits over time.

As authors, we recommended to consult with a reputable carbon offsetting organization or project developer to obtain up-to-date information on the cost of carbon credits associated with specific mangrove conservation and restoration projects.

In this unit, we present a first order analysis on carbon storage potential by dominant true mangrove species in 19 blocks of Indian Sundarbans (Table 5.5) that are distributed in the North and South 24-parganas districts of West Bengal (Table 5.6). This analysis was carried out during the postmonsoon season of 2022–23 (December 2022–February 2023) using the standard methodology as highlighted in Annexure 5.1.

We observed from Table 5.6 that the average mangrove AGB value in Indian Sundarbans (covering 19 blocks) is 3.09 Mg/ha compared to the global average mangrove AGB value of 115.23 Mg/ha (Table 5.7).

The study conducted by us in 19 blocks of Indian Sundarbans highlight a total CO_2-e value of 1960.17 tha^{-1} for 380 ha sampling area that were planted during 1990 (Table 5.6). Using the same assumptions as adopted in Table 5.4, the present valuation of 32-year true mangrove species in Indian Sundarbans is $611,598.40, which is equivalent to INR 5,01,51,065 (round off). It is to be noted that in this value the Below Ground Biomass (BGB) as well as the risk-adjusted discount rate (~10%) have not been considered. We have also eliminated those mangrove patches where the CO_2 level is much high due to decomposition of fishes and other waste materials (e.g., fish landing stations like Namkhana, Kakdwip, Fraserganj etc. or shrimp farms excavated at the cost of mangroves) in the Sagar block (Figs. 5.2 and 5.3). These areas probably act as source of carbon dioxide rather than sink, and hence may generate negative NPVs that are not financially viable while estimating carbon credit.

5.2 Carbon Credit Through Mangrove Associate Species and Underlying Soil

In addition to true mangrove species, there are also associated plant species in the same ecosystem, which also play great role in terms of ecosystem services. The entire plant spectra in mangrove ecosystem can be categorized into true mangroves and mangrove associates. The species which are adapted to mangrove habitat and do not extend into other terrestrial plant communities are referred to as true mangrove species. On contrary, plants that occur in the coastal environment and often extend beyond true mangrove habitats are referred to as mangrove associate species. Figure 5.4 presents a very common example of mangrove associate species, which is a type of salt marsh grass species, *Porteresia coarctata* that are abundantly found in the intertidal zone within the tropical belt.

Table 5.5 Above Ground Biomass (AGB) range in tha^{-1} of five dominant true mangrove tree species in 19 blocks of Indian Sundarbans

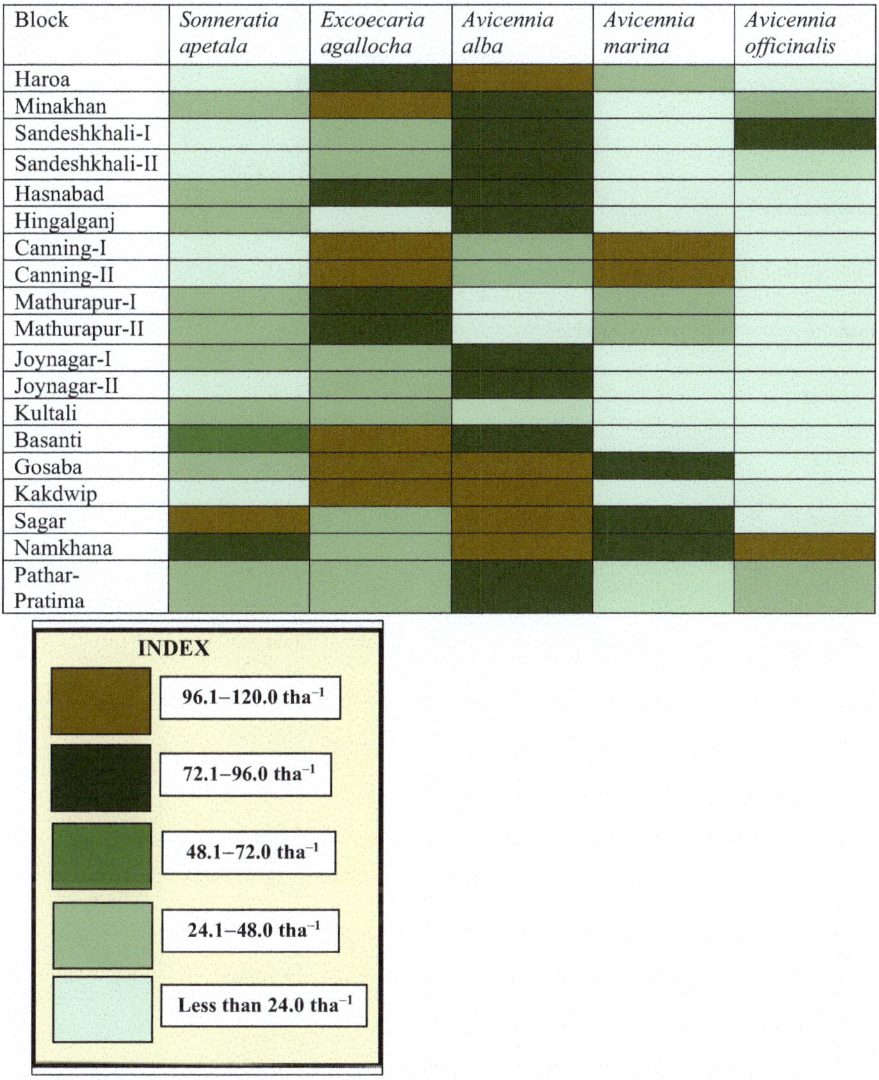

While mangrove-associate plant species can often be found in the same habitats where true mangroves thrive, they are not considered true mangroves because they lack the specific adaptations necessary to survive in the intertidal zone like pneumatophores, prop roots, stilt roots etc. or exhibit viviparous germination. They may also grow on slightly higher ground (supra-littoral zone) or on the fringes of the mangrove forest, where they are less exposed to saline water and waterlogging of the intertidal zone.

Table 5.6 CO_2-e estimated from total AGB and total AGC of dominant true mangrove species for 19 blocks in Indian Sundarbans

Block	Area (ha)	AGB* (tha^{-1})	AGC* (tha^{-1})	CO_2-e* (tha^{-1})
Haroa	15,273	45.2	20.61 (45.6%)	75.64
Minakhan	15,880	43.1	19.52 (45.3%)	71.65
Sandeshkhali-I	18,230	28.2	12.63 (44.8%)	46.37
Sandeshkhali-II	19,720	26.7	11.93 (44.7%)	43.80
Hasnabad	15,310	32.3	14.83 (45.9%)	54.41
Hingalganj	23,880	52.1	24.02 (46.1%)	88.15
Canning-I	18,790	29.8	13.05 (43.8%)	47.90
Canning-II	21,490	21.4	9.42 (44.0%)	34.56
Mathurapur-I	14,730	53.6	24.17 (45.1%)	88.72
Mathurapur-II	22,740	55.9	25.04 (44.8%)	91.91
Joynagar-I	13,100	61.4	27.57 (44.9%)	101.18
Joynagar-II	18,620	67.7	29.65 (43.8%)	108.83
Kultali	30,618	73.0	33.29 (45.6%)	122.17
Basanti	40,421	97.2	45.78 (47.1%)	168.02
Gosaba	29,673	101.5	48.01 (47.3%)	176.19
Kakdwip	25,270	73.8	33.36 (45.2%)	122.42
Sagar	28,210	117.1	52.70 (45.0%)	193.39
Namkhana	37,060	97.6	43.82 (44.9%)	160.83
Pathar Pratima	48,450	94.9	44.70 (47.1%)	164.04

*Indicates that the estimation of the values of AGB, AGC and CO_2-equivalent are based on the area of our sampling plots (20 ha in the mangrove patches per block); the estimation does not include the total area of the block

Table 5.7 Mean AGB and total AGB values of mangroves in different regions of the world

Region	Mean AGB (Mg/ha)	Mangrove area (ha)	Total AGB (Mg)	Proportion of global AGB (%)
South America	111.33 ± 58.70	20,62,231.25	22,95,94,355.48	15.12
Central America	110.29 ± 39.48	13,88,962.50	15,31,85,736.15	10.09
Southern Asia	132.60 ± 29.79	9,49,281.25	12,58,74,227.46	8.29
Western Africa	79.27 ± 34.32	14,75,343.75	11,69,44,200.85	7.7
Eastern Africa	102.79 ± 53.63	8,21,906.25	8,44,82,779.41	5.56
Caribbean	123.69 ± 31.40	5,71,493.75	7,06,90,803.56	4.65
Melanesia	149.24 ± 48.63	4,38,768.75	6,54,79,984.29	4.31
Australia and New Zealand	101.22 ± 38.71	5,23,643.75	5,30,00,621.89	3.49
Middle Africa	101.79 ± 36.88	3,93,006.25	4,00,03,008.67	2.63
Northern America	103.48 ± 44.90	3,00,956.25	3,11,42,627.18	2.05
Southern Africa	197.16 ± 52.95	43,118.75	85,01,508.15	0.56
Western Asia	134.24 ± 16.11	23,162.50	31,09,444.89	0.2
Micronesia	279.89 ± 81.67	10,162.50	28,44,427.43	0.19
Northern Africa	161.37 ± 11.47	9,481.25	15,29,943.27	0.1
Eastern Asia	114.90 ± 23.41	9,156.25	10,52,045.89	0.07
Polynesia	160.25 ± 58.88	100	16,025.34	< 0.01
Global	115.23 ± 48.89	1,30,65,675.00	1,51,87,98,427	100

Source Hu et al. (2020)

Nonetheless, these plant species are still important components of mangrove ecosystems, providing food and shelter for a variety of faunal species, and contributing to the overall biodiversity of the area.

It has been observed that mangrove associate species can also thrive luxuriantly in terrestrial region far away from mangroves. The present authors also observed the presence of *Ipomoea pes-caprae* on the road side of Kalinganagar Project Office, a Project of TATA STEEL, Odisha (Fig. 5.5). This species commonly called sand binders, are found on the sandy beaches of coastal zone or on the river mouths dominated by sand.

Ipomoea pes-caprae, also known as beach morning glory or goat's foot, is a species of flowering plant in the family Convolvulaceae. It is a vine that is found in sandy coastal areas and dunes of tropical and subtropical regions worldwide.

Fig. 5.2 Several small shrimp culture ponds have been excavated at the cost of mangroves that could otherwise be a potential sink of atmospheric carbon dioxide at local scale

Fig. 5.3 Fish landing stations established in mangrove zone after deforestation presently acts as a source of carbon dioxide rather than sink

Fig. 5.4 Mangrove associate species, *Porteresia coarctata* (commonly known as salt marsh grass); it acts as the pioneer species in the process of ecological succession

Fig. 5.5 Growth of *Ipomoea pes-caprae* along the road side of KPO under Tata Steel (Odisha)

The plant has a long, sprawling stem that can reach up to several meters in length, with large, heart-shaped leaves that are typically around 7–15 cm long. The flowers of *Ipomoea pes-caprae* are funnel-shaped, and are typically a pink or purple color, with a white or yellow throat.

Ipomoea pes-caprae is an important mangrove associate plant species that serve in stabilizing sand dunes and preventing beach erosion. Its extensive root system helps to bind sand particles together and prevent them from being blown away by the wind.

Common examples of mangrove associate species are *Porteresia coarctata, Ipomoea pes-caprae, Sesuvium portulacastrum* etc. A list of mangrove associate species commonly available in Indian Sundarbans is highlighted in Table 5.8.

We observed non-uniformity in the distribution pattern of mangrove associate species in Indian Sundarbans (Table 5.9), which is a clear indication of the growth of salinity—specific mangrove associate species in this region.

Few important ecosystem services provided by these mangrove associates are discussed in brief.

(a) **Beach stability**

The mangrove associates plant species play important roles in minimizing the action of waves, promote accretion of sediment and dune formation. All these processes increase the density of vegetation and the depth of the water. *Porteresia coarctata*

Table 5.8 Herbs, grasses and sedges associated with true mangrove floral species of Indian Sundarbans

Species	Family	Species	Family
Aeluropus logopoides	Poaceae	*Myriostachya wightiana*	Poaceae
Aerva lanata	Amaranthaceae	*Panicum repens*	Poaceae
Ammania baccifera	Lythraceae	*Paspalum vaginatum*	Poaceae
Caesalpinia crista	Caesalpiniaceae	*Phragmites karka*	Poaceae
Canavalia cathartica	Caesalpiniaceae	*Porteresia coarctata*	Poaceae
Cyperus exaltatus	Cyperaceae	*Salicornia brachiata*	Chenopodiaceae
Fimbristylis halophila	Cyperaceae	*Sacobolus carinatus*	Asclepiadaceae
F. sub-bispicata	Cyperaceae	*Scirpus triquetra*	Cyperaceae
Heliotropium curassavicm	Boraginaceae	*Sesuvium portulacastrum*	Alizoaceae
Hoya parasitica	Asclepiadaceae	*Suaeda maritima*	Chenopodiaceae
Hydrophylax maritima	Rubiaceae	*S. nudiflora*	Chenopodiaceae
Ipomoea pes-caprae	Convolvulaceae	*Viscum orientale*	Loranthaceae
Lersia hemandra	Poaceae		

Table 5.9 Check list of mangroves and mangrove associated plant species of Indian Sundarbans

Name of the family	Name of the species	W	C	E
Meliaceae	*Aglaia cucullata*	–	–	–
Rubiaceae	*Scyphiphora hydrophyllacea*	–	–	+
Acanthaceae	*Acanthus ilicifolius*	+	+	+
	Acanthus volubilis	–	+	–
Fabaceae	*Cynometra ramiflora*	–	–	–
	D. scandens	–	–	+
	D. trifoliata	–	+	+
	Dalbergia spinosa	–	–	+
Asclepiadaceae	*Sarcolobus globosus*	–	+	+
	S. carinatus	–	+	+
	Pentatropis capensis	+	+	+
Verbenaceae	*Clerodendrum inerme*	+	+	+
	C. neriifolium	–	–	+
	Premna corymbosa	–	–	–
Chenopodiaceae	*Suaeda nudiflora*	–	+	+
	S. maritima	–	+	+
	Salicornia brachiata	–	–	+
Aizoaceae	*Sesuvium portulacastrum*	–	–	+
	Trianthema portulacastrum	+	+	+
	T. triquetra	+	+	+
Ruppiaceae	*Ruppia maritima*	+	+	–
Poaceae	*Porteresia coarctata*	+	+	+
	Myriostachya wightiana	–	–	–
	Phragmites karka	+	+	–
Hippocrateaceae	*Salacia chinensis*	–	–	–
Cyperacea	*Scirpus littoralis*	+	+	–
Apocynaceae	*Cerbera odollam*	–	–	–
Sapindaceae	*Dodonaea viscosa*	–	–	–
Cluciaceae	*Calophyllum inophyllum*	–	–	–
Pteridanceae	*Acrostichum aureum*	+	+	+

W Western Indian Sundarbans; *C* Central Indian Sundarbans; *E* Eastern Indian Sundarbans

belonging to family Poaceae is a pioneer species in the sequence of ecological succession during island formation in Indian Sundarbans (Fig. 5.6). Similarly, *Ipomoea pes-caprae* with very long stem and root systems act as sediment binders and stabilize the sand dunes on the coastal zone (Fig. 5.7). The mangrove associate species trap the sediments by their long root system and thus serve as land expanders.

Fig. 5.6 *Porteresia coarctata* bed on the mudflats of Indian Sundarbans

Fig. 5.7 *Ipomoea pes-caprae*—a common sand binder mostly found in the supra-littoral zone dominated by sand

(b) Bioremediation

Mangrove ecosystems have survived the onslaught of man by using them as waste-dumping sites. The potential of mangrove associate species as agents of bioremediation is basically because of three reasons (i) flow through the floral bed disperses wastes from a point source over vast areas, (ii) the floral community filters nutrients and absorbs heavy metals from the water and sediment, (iii) the substratum/soil microbes attached to the root system absorb large number of pollutants. Researchers have documented that *Porteresia coarctata* and *Ipomoea pes-caprae* can accumulate considerable amount of Fe, Zn, Cu, Pb and even Hg in their body tissues (Mitra et al. 2014).

Recent studies carried out by our team members during January 2023 revealed the presence of considerable amount of heavy metals in the vegetative parts of these mangrove associate species and the order of accumulation is stem > leaf > root in case of *Ipomoea pes-caprae* (Tables 5.10, 5.11 and 5.12). In case of *Porteresia coarctata,* highest concentrations of Zn were observed in the vegetative parts followed by Cu and Pb (Tables 5.13, 5.14 and 5.15).

(c) Carbon Storage

The role of mangrove associate species in carbon sequestration has not been investigated in details like mangroves, although they constitute a considerable biomass in the mangrove habitats of the world.

P. coarctata is a perennial halophytic mangroves associate species of Poaceae family (Fig. 5.8). The species acts as the pioneer flora in the process of ecological succession in the marine and estuarine island ecosystem of India (Jagtap 1985; Chaudhuri and Choudhury 1994; Mitra and Banerjee 2005).

Table 5.10 Bioaccumulation of Zn in the vegetative parts of *Ipomoea pes-caprae* sampled during January 2023 from sampling stations in and around Indian Sundarbans

Station	Root	Stem	Leaf
Canning	75.04	65	37.77
Gosaba	73.05	58.34	37.53
Diamond Harbour	98.11	88.26	52.2
Nayachar Island	111.16	94.13	64.48
Kakdwip	95.22	87.16	39.99
Chemaguri	74.14	61.16	51.56
Sagar South	93.18	85.28	60.56
Jambu Island	86.04	79.38	48.4
Frasergunge	81.32	78.12	45.11
Digha	80.99	77.09	39.55
Bali	45.46	41.01	27.83
Bagmara	37.22	29.14	26.32

Table 5.11 Bioaccumulation of Cu in the vegetative parts of *Ipomoea pes-caprae* sampled during January 2023 from sampling stations in and around Indian Sundarbans

Station	Root	Stem	Leaf
Canning	33.95	30.41	23.39
Gosaba	29.9	28.6	21.74
Diamond Harbour	44.29	37.84	31.41
Nayachar Island	49.16	44.04	33.44
Kakdwip	43.32	35.36	28.76
Chemaguri	32.14	30.09	22.64
Sagar South	40.25	35.38	28.26
Jambu Island	39.9	32.39	25.6
Frasergunge	36.57	31.95	25.98
Digha	35.85	31.06	23.75
Bali	28.2	25.27	20.89
Bagmara	26.22	22.03	20.99

Table 5.12 Bioaccumulation of Pb in the vegetative parts of *Ipomoea pes-caprae* sampled during January 2023 from sampling stations in and around Indian Sundarbans

Station	Root	Stem	Leaf
Canning	15.19	12.14	11.99
Gosaba	18.28	18.01	15.09
Diamond Harbour	13.08	8.65	8.31
Nayachar Island	14.69	9.95	7.88
Kakdwip	22.49	20.97	13.42
Chemaguri	15.97	13.1	10.28
Sagar South	19.19	19.5	16.08
Jambu Island	18.43	16.1	13.89
Frasergunge	19.05	17.09	15.11
Digha	25.89	24.08	17.76
Bali	14.09	12.99	11.89
Bagmara	11.22	6.67	3.07

We monitored the stored carbon in the above and below ground biomass of *Porteresia coarctata* for 12 consecutive years (April 2011–April 2022) along the Hooghly estuary in the maritime State of West Bengal, India.

Two sampling stations were selected along the Hooghly estuary in the northeast coast of India, namely Sagar Island (21° 37′ 49.90″ N and 88° 04′ 00.51″ E) and Kakdwip (21° 52′ 26.50″ N and 88° 08′ 04.48″ E) as shown in Fig. 5.9.

Sagar Island is located at the confluence of the Hooghly estuary and the Bay of Bengal in the downstream region, and Kakdwip is in the upstream region, where the salinity is relatively low compared to Sagar Island (Banerjee et al. 2013, 2017; Zaman et al. 2013; Trivedi et al. 2016; Mitra 2018; Dutta et al. 2020; Dhar et al. 2021).

Table 5.13 Bioaccumulation of Zn in the vegetative parts of *Porteresia coarctata* sampled during January 2023 from sampling stations in and around Indian Sundarbans

Station	Root	Stem	Leaf
Canning	100.49	88.27	62.13
Gosaba	95.06	80.72	61.2
Diamond Harbour	121.76	110.69	76.73
Nayachar Island	133.96	117.23	86.67
Kakdwip	119.84	110.6	61.76
Chemaguri	97.21	83.74	75.37
Sagar South	116.62	108.37	73.5
Jambu Island	110.62	102.74	70.71
Frasergunge	103.61	101.26	67.38
Digha	102.06	101.04	62.76
Bali	69.62	64.36	52.62
Bagmara	58.85	52.67	51.43

Table 5.14 Bioaccumulation of Cu in the vegetative parts of *Porteresia coarctata* sampled during January 2023 from sampling stations in and around Indian Sundarbans

Station	Root	Stem	Leaf
Canning	48.18	44.62	37.55
Gosaba	43.83	42.92	35.67
Diamond Harbour	58.53	52.56	44.53
Nayachar Island	61.88	57.76	47.68
Kakdwip	58.59	50.59	42.09
Chemaguri	45.23	44.21	38.63
Sagar South	54.65	48.87	41.79
Jambu Island	53.79	46.51	40.08
Frasergunge	50.89	45.87	39.81
Digha	49.64	45.09	38.48
Bali	42.21	39.19	35.22
Bagmara	39.19	36.48	34.74

Yearly samplings for biomass estimation of *P. coarctata* were carried out at ebb from March 2011 to March 2022 in the intertidal mudflats along the Hooghly estuary (in the northeast coast of India). A perpendicular belt (~50 m width) transect to the shore was laid on the *Porteresia* sp. bed from low tide to high tide mark at each station. The length of transect varied from 20 to 250 m depending upon the horizontal extent of bed in the mudflats. At each station, two to three perpendicular transects were laid down, depending upon the extent of bed parallel to the mudflat. The biomass was estimated by removing all the plants along with roots, and shoots from five quadrants (each of 1 m^2 size) for each station. Plant material was thoroughly washed in the ambient water immediately after collection, as well as with distilled water, to remove adhering debris and sediments. The various vegetative parts (above and below

Table 5.15 Bioaccumulation of Pb in the vegetative parts of *Porteresia coarctata* sampled during January 2023 from sampling stations in and around Indian Sundarbans

Station	Root	Stem	Leaf
Canning	24.07	31.03	30.88
Gosaba	24.17	26.9	23.97
Diamond Harbour	31.97	27.54	27.2
Nayachar Island	23.58	18.8	17.77
Kakdwip	21.38	19.88	18.32
Chemaguri	24.86	23.99	23.17
Sagar South	28.05	28.39	24.97
Jambu Island	27.32	27.99	22.78
Frasergunge	17.93	13.88	12.32
Digha	24.78	22.94	16.65
Bali	22.97	21.88	18.78
Bagmara	13.21	11.44	9.89

Fig. 5.8 *Porteresia coarctata*—abundantly available on the intertidal mudflats of Indian Sundarbans

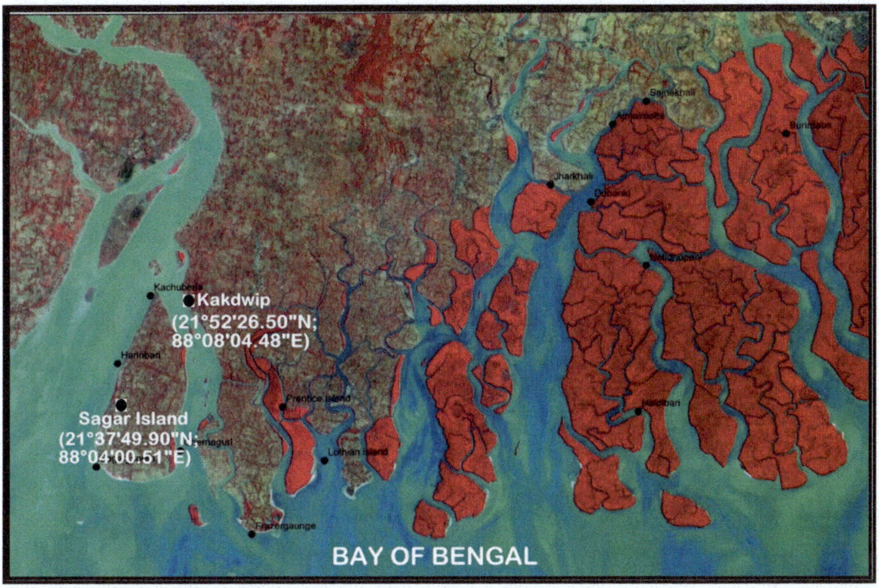

Fig. 5.9 Sampling sites along the Hooghly estuary selected for the present study

ground) were separated, oven dried and weighed, and the results were expressed as gm m^{-2} on an average basis.

Direct estimations of percent carbon in the AGB and BGB were done by *Vario* MACRO Elementar CHN analyzer, after grinding the oven-dried above ground and below ground structures separately. This exercise was performed for the samples of both the stations.

We observed significant role of salinity in the carbon storage potential of the species as the values of stored carbon was consistently higher in the low saline region like Kakdwip compared to Sagar Island. Thus, site-wise monitoring of stored carbon is needed to obtain the ground-zero reality.

In Sagar Island, the AGB ranged from 131.24 to 283.19 gm m^{-2}, whereas in Kakdwip, the values were relatively high ranging from 156.81 to 351.44 gm m^{-2}. The BGB values in Sagar Island ranged between 146.14 to 301.55 gm m^{-2}. In Kakdwip, the BGB values ranged from 153.17 to 368.76 gm m^{-2}.

Figures 5.10 and 5.11 represent the AGB and BGB of the species collected from the two selected stations.

The stored carbon in the AGB (referred to as AGC) ranged from 50.87 to 105.61 gm m^{-2} in Sagar Island, whereas in Kakdwip the values ranged from 63.86 to 145.60 gm m^{-2}. In Sagar Island, the BGC ranged from 66.76 to 136.70 gm m^{-2}. The values of BGC in *P. coarctata* sampled from the hyposaline Kakdwip station ranged from 72.30 to 176.05 gm m^{-2} (Figs. 5.12 and 5.13).

We observed significant spatio-temporal variations in AGB, BGB, AGC and BGC of the species ($p < 0.01$) as shown in Table 5.16.

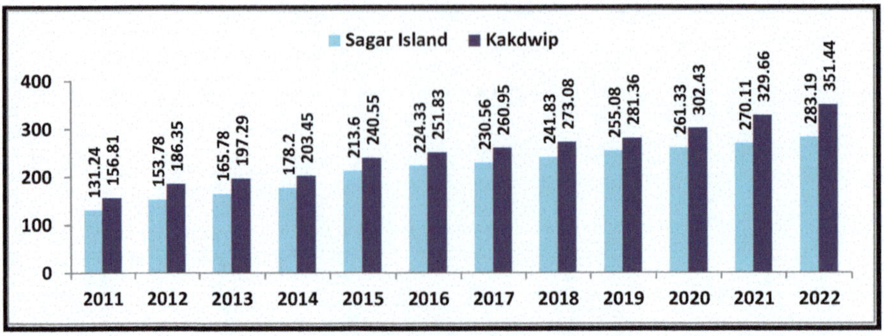

Fig. 5.10 AGB of *P. coarctata* samples of Sagar Island and Kakdwip

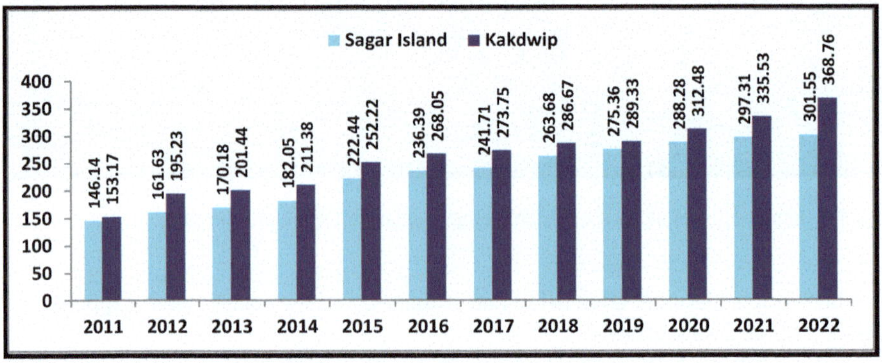

Fig. 5.11 BGB of *P. coarctata* samples of Sagar Island and Kakdwip

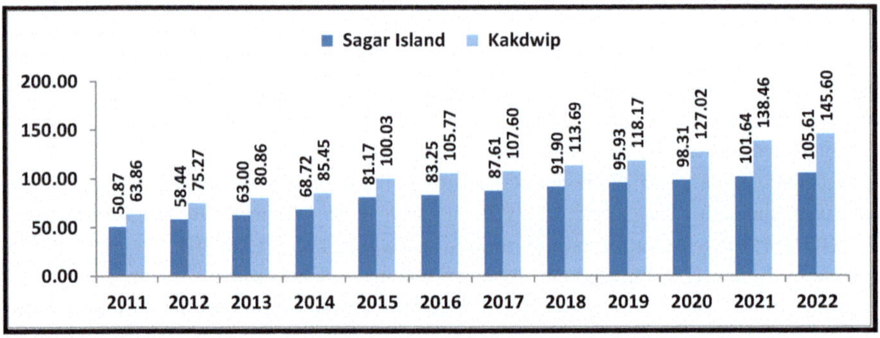

Fig. 5.12 AGC of *P. coarctata* samples of Sagar Island and Kakdwip

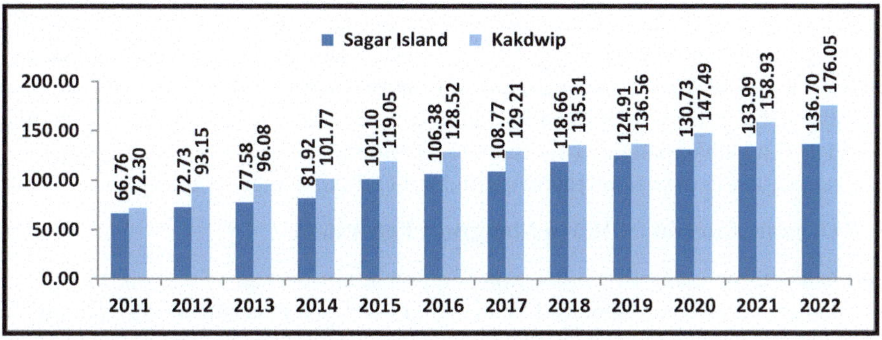

Fig. 5.13 BGC of *P. coarctata* samples of Sagar Island and Kakdwip

Table 5.16 Spatio-temporal variations of AGB, BGB, AGC and BGC of *P. coarctata* based on the field data from 2011 to 2022

Variables	F_{cal}	F_{crit}
1. AGB		
Between years	59.32	2.81
Between stations	76.48	4.84
2. BGB		
Between years	65.20	2.81
Between stations	51.10	4.84
3. AGC		
Between years	27.66	2.81
Between stations	93.33	4.84
4. BGC		
Between years	45.98	2.81
Between stations	70.97	4.84

Many literatures are available on the stored carbon in the coastal vegetation (Twilley et al. 1992; Connor et al. 2001; Drake et al. 2015; Fourqurean et al. 2012; Chastain et al. 2018). However, stored carbon in marshy vegetation in the coastal areas of West Bengal has not been explored in detail preferably with respect to time series frame. *P. coarctata* is profusely available in the estuarine water as well as saline habitats of coastal West Bengal and therefore, has the capacity to sequester carbon in a wide range of salinity. Based on this, the present research was undertaken continuously for 12 years (2011–2022) to assess the stored carbon in *P. coarctata* sampled from two different sites with contrasting salinity. Sagar Island has high salinity due to its location in the downstream region at the mixing zone of the River Hooghly and the Bay of Bengal and Kakdwip has relatively low salinity due to its location in the upstream region along the Hooghly estuary. Considerable variations in AGB, BGB, AGC and BGC observed in the present study may be attributed to such variation in salinity profile.

In conclusion, it can be advocated that the biomass and carbon storage capacity of *P. coarctata* vary with spatial locations, mostly regulated by sea water inflow and dilution factor. Effective management of dilution factor through adequate input of freshwater by gravity flow from the rainwater harvested ponds in the hypersaline mudflats is a viable roadmap to accelerate the biomass of the saltmarsh grass species that subsequently determine the magnitude of carbon storage by the species.

Soil Organic Carbon (SOC) in Mangrove Ecosystem

Soil beneath the mangrove vegetation is a potential reservoir of carbon. Mangrove and mangrove associate species like marsh grasses, sea grass meadows accumulate soil particles and contribute to the reservoir of Soil Organic Carbon (SOC). The decomposition of mangrove and associated species also enriches the soil with SOC through microbial degradation of litter and detritus sourced from halophytes. The organic matter in the mangrove soil extends to many meters depth and provide long-term storage of SOC. The mangrove habitats in the world occupy very small area, but are considered as the major contributors of carbon to marine sediments (Duarte et al. 2013). In the domain of carbon sequestration, the mangrove ecosystems are of particular interest as they store and sequester carbon both in the biomass and underlying soil (Donato et al. 2011; Ezcurra et al. 2016; Almahasheer et al. 2017; Kauffman et al. 2017).

According to Donato et al. (2011) the storage volume of SOC in mangrove is about five times as stored in tropical upland forest. The reason behind such high volume of SOC in mangrove soil may be attributed to two important factors viz (i) high productivity of mangrove biomass and (ii) slow decomposition rate of mangrove soil (Alongi 2012).

In addition to these factors, the complicated root structure of mangroves bind the soil particles rich in carbon from the ambient aquatic phase. According to McKee et al. (2007) the root system of mangroves trap allochthonous organic material above the carbon rich peat composed of dead root materials and this can extend up to 10 m depth. Cooray et al. (2021) observed that SOC in mangrove habitat comprise mostly of mangrove biomass, which is about 90% of the total carbon stock in the mangrove soil. Such large dimension of stored mangroves both in the biomass and underlying soil of mangrove vegetation has sparked the interest of many companies related to carbon offset programmes (Donato et al. 2011; Forquerrean et al. 2012).

We carried out a SOC monitoring program at 14 stations in Indian Sundarbans during 2018–2022 considering a depth of 10 cm from the surface. The sampling was conducted during the low tide phase in December in every year and activities of each of the selected stations were identified to evaluate their contribution as drivers to SOC stock (Table 5.17; Fig. 5.14).

The results of our long-term analysis are presented in Figs. 5.15, 5.16, 5.17, 5.18 and 5.19.

The significant variation ($p < 0.0001$) of soil organic carbon between anthropogenically stressed western zone and non-disturbed central and eastern zones (ANOVA carried out with the values of Fig. 5.20) may be attributed to a large extent by human activities, mangrove floral richness, and physical factors like accretion

Table 5.17 Major drivers influencing the SOC pool in Indian Sundarbans

Station with Stn. No. in bracket ()	Longitude and latitude	Major activity	Magnitude
Kachuberia (1)	88° 08′ 04.43″ E; 21° 52′ 26.50″ N	Navigational channel	+++
		Passenger vessel jetties	+++
		Shrimp culture farms	+
		Market place	++
Harinbari (2)	88° 04′ 52.98″ E; 21° 47′01.36″ N	Mangrove patches	++
		Unorganized fishing activities	+
Sagar South (3)	88° 03′ 06.17″ E; 21° 38′ 54.37″ N	Pilgrims	+++
		Tourism	+++
		Navigational channel	+++
		Erosion (sea facing)	+++
		Mangrove patches	++
Chemaguri (4)	88° 10′ 07.03″ E; 21° 39′ 58.15″ N	Fish landing stations	+++
		Tourism	+++
		Mangrove patches	++
		Shrimp culture farms	++
Fraserganj (5)	88° 15′ 15.63″ E; 21° 33′ 11.84″ N	Shrimp culture farms	++
		Mangrove forest	+++
		Fish landing stations	+
		Market place	++
Prentice Island (6)	88° 17′ 10.04″ E; 21° 42′ 40.97″ N	Mangrove patches	+++
Lothian island (7)	88° 22′ 13.99″ E; 21° 39′ 01.58″ N	Mangrove forest (protected area)	+++
		Navigational channel	+++
		Erosion	+++
		Wave action	+++

(continued)

Table 5.17 (continued)

Station with Stn. No. in bracket ()	Longitude and latitude	Major activity	Magnitude
Sajnekhali (8)	88° 46′ 10.08″ E; 22° 05′ 13.04″ N	Mangrove forest	+++
		Tourism	++
Amlamethi (9)	88° 44′ 26.07″ E; 22° 03′ 54.02″ N	Mangrove forest	+++
		Tourism (occasional)	+
		Shrimp and prawn culture farms	+
Jharkhali (10)	88° 41′ 47.25″ E; 22° 05′ 52.82″ N	Mangrove forest	+++
Dobanki (11)	88° 45′ 20.06″ E; 21° 59′ 24.04″ N	Accretion zone	++
		Mangrove forest	+++
Netidhopani (12)	88° 44′ 39.4″ E; 21° 55′ 14.9″ N	Mangrove forest	+++
Haldibari (13)	88° 46′ 44.9″ E; 21° 43′ 01.4″ N	Accretion zone	++
		Mangrove forest	+++
Burirdabri (14)	89° 01′ 43.6″ E; 22° 04′ 39.2″ N	Mangrove forest (protected area)	+++

+, ++, and +++ indicate low, medium, and high magnitude respectively for the major activities in the selected stations; n and AGMB represent number of mangrove species and above ground mangrove biomass (t ha^{-1}) of three dominant species respectively

Data sources Chaudhuri and Choudhury (1994), Mitra et al. (2010), Mitra et al. (2011), Mitra (2020), Mitra et al. (2022)

and erosion. Anthropogenic activities like fish landing, tourism and unplanned urban development and shrimp farms contribute appreciable amount of organic load in stations like Kachuberia (Stn. 1) and Fraserganj (Stn. 5). The presence of shrimp farms at Chemaguri (Stn. 6) along with 22 years old mangrove vegetation (17 species) may be attributed to highest organic carbon level in the surface soil. The western Indian Sundarbans (encompassing stations 1–7) is under severe stress due to intense industrialization, rapid urbanization and unplanned tourism and aquaculture activities, which contribute appreciable organic carbon in the soil. The relatively low organic carbon at Sagar South (Stn. 3) is due to its location at sea front where wave action and tidal amplitude is maximum (range 3.0–5.0 m and mean = 3.5 m). Continuous erosion of this island may be the reason behind minimum retention of organic matter in the intertidal zone. The low organic carbon at stations 8–14 confirms the anthropogenic origin of organic load, which is almost absent in these stations (control zone).

It is to be noted that the values of SOC exhibited a gradual rise with time except at station 3, which is an erosion prone zone (Fig. 5.20).

The present study is significant from the point that the area has not yet witnessed the light of documentation of SOC stock although Above Ground Mangrove Biomass

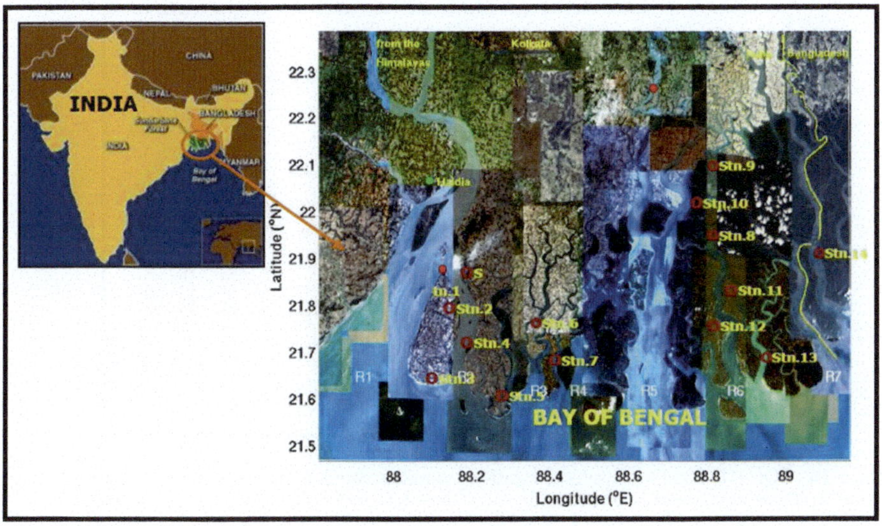

Fig. 5.14 Map of the study region showing the sampling stations (Stn.). R1–R7 are the seven rivers of Sundarbans starting from west to east, namely Hooghly, Mooriganga, Saptamukhi, Thakuran, Matla, Gosaba and Harinbhanga

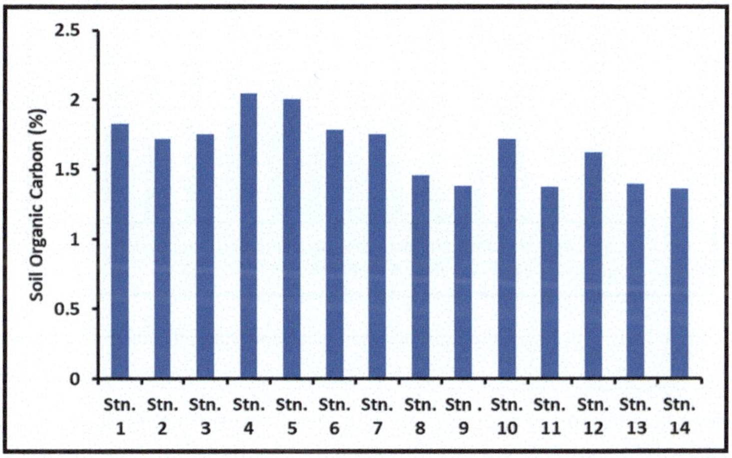

Fig. 5.15 SOC (%) level in 14 selected stations in Indian Sundarbans during postmonsoon of 2018

(AGMB) and carbon storage have been studied by several workers (Mitra et al. 2009, 2010, 2011; Raha et al. 2013; Sengupta et al. 2013; Bhattacharyya et al. 2015; Mitra and Gati 2015; Agarwal et al. 2016, 2017; Banerjee et al. 2016; Chakraborty et al. 2016). Donato et al. (2011) quantified whole-ecosystem C storage in mangroves across a broad tract of the Indo-Pacific region, which includes the Bangladesh Sundarbans. The study, however, did not cover the lower Gangetic soil sustaining 38% of

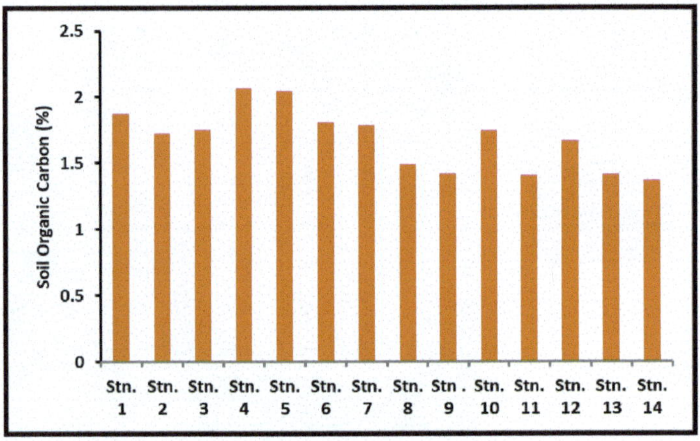

Fig. 5.16 SOC (%) level in 14 selected stations in Indian Sundarbans during postmonsoon of 2019

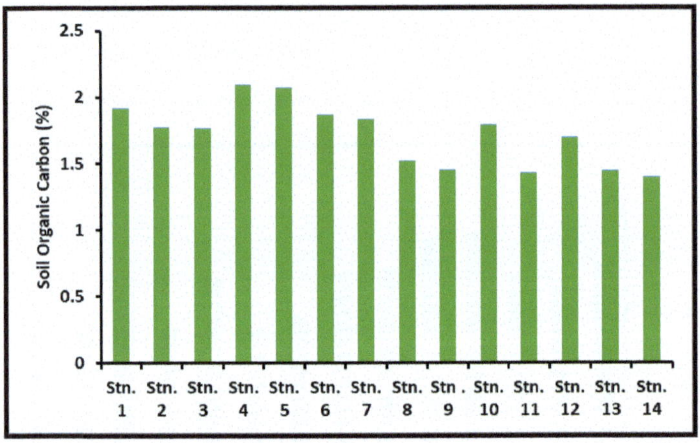

Fig. 5.17 SOC (%) level in 14 selected stations in Indian Sundarbans during postmonsoon of 2020

the total Sundarbans in the Indian part. The present approach is thus an attempt to fill this gap area and establish a continuous five-year baseline data of soil carbon in the mangrove dominated Indian part of Sundarbans.

5.3 Gap Area in Blue Carbon Credit

Blue carbon credits are a mechanism for financing the conservation and restoration of coastal and marine ecosystems that sequester and store carbon, such as mangroves, seagrasses, and salt marshes. These credits are like carbon credits in other sectors,

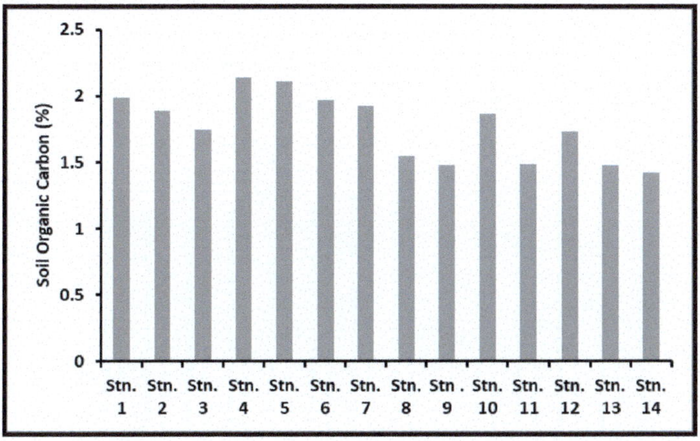

Fig. 5.18 SOC (%) level in 14 selected stations in Indian Sundarbans during postmonsoon of 2021

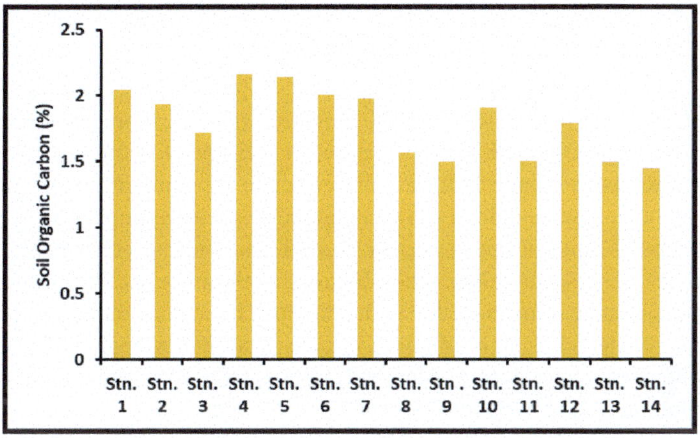

Fig. 5.19 SOC (%) level in 14 selected stations in Indian Sundarbans during postmonsoon of 2022

such as forestry and agriculture, and represent a quantified and verified amount of carbon dioxide equivalent (CO_2-e) that has been removed or minimized through the conservation or restoration of blue carbon ecosystems.

Blue carbon credits can be sold to companies or individuals who wish to offset their carbon emissions and meet their carbon reduction targets. The revenue generated from the sale of blue carbon credits can be used to finance conservation and restoration projects, provide incentives to local communities to protect coastal ecosystems, and support sustainable livelihoods.

There are several initiatives and programs that facilitate the creation and trading of blue carbon credits, such as the Blue Carbon Initiative, the Verified Carbon Standard's Blue Carbon methodology, and the Carbon Credit and Investment Framework for

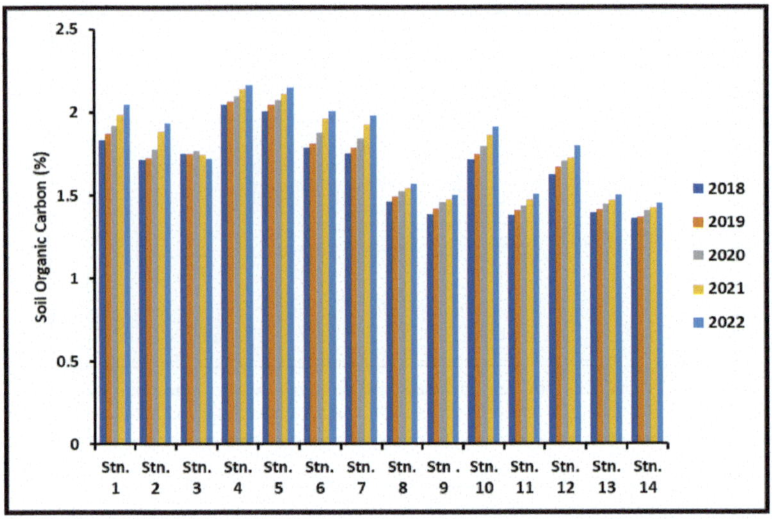

Fig. 5.20 SOC (%) level in 14 selected stations during postmonsoon of 5 consecutive years (2018–2022)

Mangrove Conservation and Restoration. However, the development of blue carbon credit projects requires careful consideration of a range of ecological, social, and governance factors to ensure that they are effective in achieving their carbon and conservation objectives and do not have unintended negative impacts.

Blue carbon is the carbon stored in coastal and marine ecosystems, such as mangroves, seagrasses, and salt marshes. Estimating the amount of blue carbon stored in these ecosystems is important for understanding their contribution to mitigating climate change and for designing effective conservation and restoration strategies. However, there are some gaps in blue carbon estimation in coastal vegetation that are still being addressed by researchers and practitioners.

1. **Limited access to data**: One of the main challenges in estimating blue carbon in coastal vegetation is the availability of data on ecosystem carbon stocks and fluxes, which is very minimum and mostly classified. This is particularly true for the rigorous environment like Sundarbans, where field measurements are difficult to obtain particularly during monsoon. Moreover, the presence of tigers (*Panthera tigris tigris*) in many islands of Sundarbans makes the field sampling extremely difficult. To address this issue, researchers are developing remote sensing and modeling approaches to estimate carbon stocks and fluxes in these ecosystems. But what about ground truth verification?
2. **Variation in blue carbon stock magnitude**: An important blockage in the domain of blue carbon estimation is the high spatial and temporal variability in ecosystem carbon stocks, which are functions of factors such as hydrological parameters, sedimentation, and nutrient availability. These factors are highly dynamic and governed mostly by tides. This causes great difficulty to accurately

estimate carbon stocks and fluxes and requires the use of multiple methods and data sources to obtain a comprehensive understanding of the ecosystem's carbon dynamics. Multidisciplinary approaches are also lacking to remove this blockage.

3. **Lack of standard estimation method**: Lack of standardization in blue carbon estimation methods and protocols is another challenge in this field. Different studies use different methods and assumptions, making it difficult to compare results and draw meaningful conclusions. To address this issue, the Blue Carbon Initiative and other organizations are working to develop standardized methods and protocols for blue carbon estimation. We can highlight the case of estimating Above Ground Biomass (AGB) in this context where different formulae are used to ascertain the tree volume.
4. **Limited understanding of coastal ecosystem dynamics**: There is still a gap in understanding the regulatory role of different environmental parameters on carbon storage by mangroves, salt marshes, sea grasses and other halophytes. The role of different species and their interactions in carbon cycling is not yet fully understood, and the effects of disturbance and management interventions on ecosystem carbon stocks and fluxes are not accurately quantified. Further research is needed to address these gaps in knowledge and improve our understanding of blue carbon in coastal vegetation.
5. **Knowledge and data gap on phytoplankton community**: Phytoplankton, being the key players of 'biological pump' have not yet been fully researched on their carbon storage potential and their volume in relation to hydrological parameters. With their position at the base of marine and estuarine food web, their role in mitigating carbon dioxide level needs to be evaluated in details.
6. **Lack of standard forecasting model**: Natural disasters like cyclones, tornadoes, hurricanes etc. pose extreme adverse impacts on blue carbon community. Erosion of intertidal mudflats that primarily sustain the blue carbon community is also associated with natural disasters. Although several computer programs have been developed by the present team members of the present authors (vide Annexure 5.2) to evaluate biodiversity of blue carbon community, but very few AI—based forecasting models like Neural AutoRegression (NAR) are available that can forecast the hydrological parameters or blue carbon biomass growth based on past data bank. This often creates uncertainty in case ROI in the domain of blue carbon.

5.4 Take Home Messages

[A] Mangrove vegetations comprising both true mangroves and associate species play an important role in mitigating climate change through carbon sequestration. These halophytes absorb and store carbon dioxide from the atmosphere in their biomass and in the soil/sediment beneath them. This makes the mangrove ecosystem a potential natural carbon sink and a unique reservoir to gain carbon credit. Carbon credits are tradeable permits that represent a certain amount of carbon dioxide or other greenhouse gas emissions that have been prevented, reduced, or removed from the atmosphere. Companies or individuals can purchase these credits to offset their own carbon emissions, contributing to the global effort to mitigate climate change.

[B] Mangrove conservation and restoration projects can generate carbon credits through quantification of the amount of carbon stored in the mangrove ecosystem. This involves measuring the carbon content of the mangrove trees, the sediment, and the associated fauna and flora, and using established methodologies to calculate the amount of carbon dioxide equivalent that has been sequestered. These carbon credits can then be sold to companies or individuals who want to offset their carbon emissions. The revenue generated from the sale of these credits can be used to finance further mangrove conservation and restoration efforts by the local communities, creating a positive feedback loop that benefits both the environment and the local communities, preferably the coastal population or the people living in the delta complex and estuarine region. Thus, carbon credits through mangrove plants offer a promising avenue for mitigating climate change while promoting sustainable development in coastal regions.

[C] There are several initiatives and programs that facilitate the creation and trading of blue carbon credits, such as the Blue Carbon Initiative, the Verified Carbon Standard's Blue Carbon methodology, and the Carbon Credit and Investment Framework for Mangrove Conservation and Restoration. However, the development of blue carbon credit projects requires careful consideration of a range of ecological, social, and governance factors to ensure that they are effective in achieving their carbon and conservation objectives and do not have any types of unintended negative impacts.

[D] Gaps existing in the domain of blue carbon credit has created uncertainty to standardise the price of blue carbon. These gaps include limited access to data on blue carbon biomass and stored carbon, lack of standard estimation methodology of blue carbon, limited understanding on coastal ecosystem dynamics, lack of AI-based standard forecasting model on the growth in biomass and stored carbon in halophytes and limited understanding on the role of marine and estuarine phytoplankton in sequestering carbon dioxide.

Annexure 5.1: Methodology for AGB and AGC Estimation of True Mangrove Tree Species

A. Above Ground Biomass (AGB) Estimation

Above Ground Biomass (AGB) of tree species refers to the sum of stem, branch and leaf biomass that are exposed above the soil.

i. **Stem Biomass Estimation**

The stem volume of each species in each plot (10 m × 10 m) was estimated using the Newton's formula (Husch et al. 1982).

$$V = h/6(A_b + 4A_m + A_t)$$

where V is the volume (in m^3), h is the height measured with laser beam (BOSCH DLE 70 Professional model), and A_b, A_m, and A_t are the areas at base, middle and top respectively. Specific gravity (G) of the wood was estimated taking the stem cores by boring 4.5 cm deep and compared with the standard data of FAO (https://www.fao.org/3/w4095e/w4095e0c.htm). This was converted into stem biomass (B_S) as per the expression $B_S = GV$. The stem biomass of individual tree was finally multiplied by the number of trees of each species in all the selected plots and the mean values are expressed in t ha^{-1} (Fig. 5.21).

In this study, aerial images were taken by a drone camera (Phantom-3 Professional, Djibouti) which has four propellers, a camera, a GPS (Global Positioning System) receiver, and a gimbal. Further, it has an exclusive remote controller. The camera used for the experiment can take 1.2 M-pixel images and video with 4 K (3840 × 2160) images. We used this parallel system to estimate the exact height of the trees in meters, number of branches, and number of leaves in each branch (Fig. 5.22).

Fig. 5.21 DBH measurement at 1.3 m height from the ground for estimation of the biomass of the tree

Fig. 5.22 Capture of aerial image by drone to know the height, number of branches and leaves of the trees

ii. **Branch Biomass Estimation**

The total number of branches irrespective of size was counted on each of the sample trees. These branches were categorized on the basis of basal diameter into three groups, viz. <6 cm, 6–10 cm and >10 cm. The leaves on the branches were removed by hand. The branches were cut in to pieces and oven-dried at 70 °C overnight in hot air oven in order to remove moisture content if any present in the branches. Dry weight of two branches from each size group was recorded separately using the equation of Chidumaya (1990).

$$B_{db} = n_1bw_1 + n_2bw_2 + n_3bw_3 = \sum n_ibw_i$$

where B_{db} is the dry branch biomass per tree, n_i the number of branches in the *i*th branch group, b_{wi} the average weight of branches in the *i*th group and i = 1, 2, 3, … n are the branch groups. The mean branch biomass of individual tree was finally multiplied with the number of trees of each species in all the plots for each site and expressed in t ha^{-1}.

iii. **Leaf Biomass Estimation**

For leaf biomass estimation, one tree of each species per plot was randomly considered. All leaves from nine branches (three of each size group) of individual trees of each species were removed and oven dried at 70 °C and dry weight (species-wise) was estimated. The leaf biomass of each tree was then calculated by multiplying the average biomass of the leaves per branch with the number of branches in that tree. Finally, the dry leaf biomass of the selected species (for each plot) was recorded as per the expression:

$$L_{db} = n_1 Lw_1 N_1 + n_2 Lw_2 N_2 + \cdots n_i Lw_i N_i$$

where L_{db} is the dry leaf biomass of selected urban species per plot, $n_1 \ldots n_i$ are the number of branches of each tree of the species, $Lw_1 \ldots Lw_i$ are the average dry weight of leaves removed from the branches and $N1 \ldots N_i$ are the number of trees per species in the plots. This exercise was performed for all the sites and the mean results were finally expressed in t ha^{-1}.

B. **Above Ground Carbon (AGC) Estimation of Trees**

Direct estimation of percent carbon in the AGB (referred to as AGC) was done by CHN analyzer, after grinding and random mixing the oven-dried stem, branches and leaves separately for each species. For this, a portion of fresh sample of stem, branch, and leaf from trees (of each species) was oven dried at 70 °C, randomly mixed and ground to pass through a 0.5 mm screen (1.0 mm screen for leaves). The carbon content (in %) was finally analyzed for each part of each species through a *Vario MACRO elementar CHN* analyzer (Fig. 5.23).

The mean carbon values of these vegetative parts (expressed in %) were considered as the stored carbon in the AGB of each species (referred to as AGC).

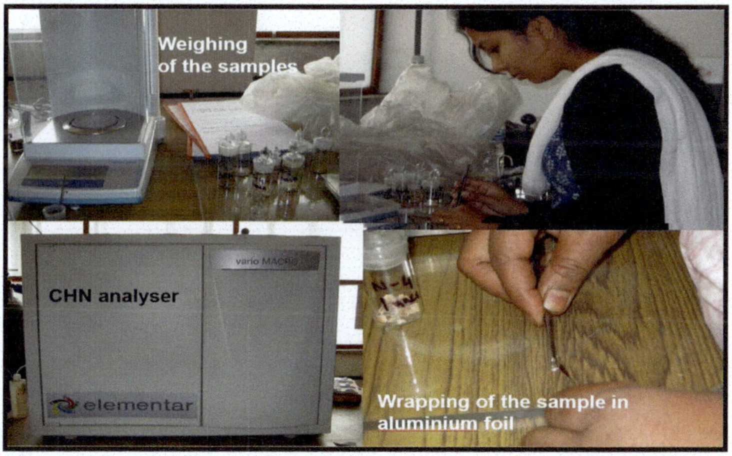

Fig. 5.23 Analysis of carbon percentage in plant samples through CHN analyzer

Annexure 5.2: Computation of Ecological Indices to Evaluate the Biodiversity Status of Phytoplankton using Python (As Example, the File Name Has Been Given Phytoplankton_abhijit and 4 Coastal Stations Have Been Selected to Feed the Data)

```
import pandas as pd
import numpy as np
import matplotlib.pyplot as plt

data1 =  pd.read_excel('phytoplankton_abhijit.xlsx',sheet_name=0)
stock1 = data1['Standing Stock (No. × 105/L)']
p1 = stock1/stock1.sum()
shannon_weiner_index_coastal station1= -np.sum(p1 * np.log(p1))
print('Shannon weiner index for coastal station 1 = ',shannon_weiner_index_coastal
station1)

data2 =  pd.read_excel('phytoplankton_abhijit.xlsx',sheet_name=1)
stock2 = data2['Standing Stock (No. × 105/L)']
p2 = stock2/stock2.sum()
shannon_weiner_index_coastal station2= -np.sum(p2 * np.log(p2))
print('Shannon weiner index for coastal station 2 = ',shannon_weiner_index_coastal
station2)

data3 =  pd.read_excel('phytoplankton_abhijit.xlsx',sheet_name=2)
stock3 = data3['Standing Stock (No. × 105/L)']
p3 = stock3/stock3.sum()
shannon_weiner_index_coastal station3= -np.sum(p3 * np.log(p3))
print('Shannon weiner index for coastal station 3 = ',shannon_weiner_index_coastal
station3)

data4 =  pd.read_excel('phytoplankton_abhijit.xlsx',sheet_name=3)
stock4 = data4['Standing Stock (No. × 105/L)']
p4 = stock4/stock4.sum()
shannon_weiner_index_coastal station4= -np.sum(p4 * np.log(p4))
print('Shannon weiner index for coastal station 4 = ',shannon_weiner_index_coastal
station4)

I1 = p1**2
index_of_dominance1 = I1.sum()
print('Index of dominance for coastal station 1 = ',index_of_dominance1)

I2 = p2**2
index_of_dominance2 = I2.sum()
print('Index of dominance for coastal station 2 = ',index_of_dominance2)

I3 = p3**2
index_of_dominance3 = I3.sum()
print('Index of dominance for coastal station 3 = ',index_of_dominance3)

I4 = p4**2
index_of_dominance4 = I4.sum()
print('Index of dominance for coastal station 4 = ',index_of_dominance4)

species = data1['Species'].value_counts()
total_species = species.sum()

EI1 = shannon_weiner_index_coastal station1/np.log(total_species)
print('Evenness index for coastal station 1 = ',EI1)

EI2 = shannon_weiner_index_coastal station2/np.log(total_species)
print('Evenness index for coastal station 2 = ',EI2)

EI3 = shannon_weiner_index_coastal station3/np.log(total_species)
print('Evenness index for coastal station 3 = ',EI3)
```

```
SWI = [shannon_weiner_index_coastal station1, shannon_weiner_index_coastal
station2, shannon_weiner_index_coastal station3, shannon_weiner_index_coastal
station4]
plt.figure(figsize=(20,12))
plt.bar(coastal stations,SWI)
plt.xlabel('Coastal station')
plt.ylabel('Shannon weiner index')
plt.title('Shannon weiner index of different coastal stations')
plt.show()

IOD = [index_of_dominance1, index_of_dominance2, index_of_dominance3,index_of_domin
ance4]
plt.figure(figsize=(20,12))
plt.bar(coastal stations,IOD)
plt.xlabel('Coastal station')
plt.ylabel('Index of dominance')
plt.title('Index of dominance of different coastal stations')
plt.show()

EI = [EI1, EI2, EI3, EI4]
plt.figure(figsize=(20,12))
plt.bar(coastal stations,EI)
plt.xlabel('Coastal station')
plt.ylabel('Evenness index')
plt.title('Evenness index of different coastal stations')
plt.show()
```

References

Agarwal S, Islam S, Zaman S, Mitra A (2017) Carbon stocks in the mangrove vegetation of Lothian wild life sanctuary of Indian Sundarbans: implication for climate change mitigation. Glob J Eng Sci Res 4(10):1–9

Agarwal S, Pramanick P, Zaman S, Roy A, Mitra A (2016) Status of stored carbon in mangroves of lower Gangetic delta. Glob J Eng Sci Res 7–16

Almahasheer H, Serrano O, Duarte CM, Arias-Ortiz A, Masque P, Irigoien X (2017) Low carbon sink capacity of red sea mangroves. Sci Rep 7:9700. https://doi.org/10.1038/s41598-017-10424-9

Alongi DM (2012) Carbon sequestration in mangrove forests. Carbon Manag 3:313–322. https://doi.org/10.4155/cmt.12.20

Anonymous (2008) Mangroves for the future: national strategy and action plan, India (Revised Draft). Ministry of Environment and Forest, Government of India, pp 63

Banerjee K, Amin G, Fazli P, Pramanick P, Zaman S, Mitra A (2016) *Excoecaria agallocha*: a potential mangrove species in context to Carbon storage in high Saline zone. J Env Sci Comput Sci Eng Technol 5(2):115–120

Banerjee K, Gatti RC, Mitra A (2017) Climate change-induced salinity variation impacts on a stenoecious mangrove species in the Indian Sundarbans. Ambio (springer) 46:492–499

Banerjee K, Sengupta K, Raha AK, Mitra A (2013) Salinity based allometric equations for biomass estimation of Sundarban mangroves. Biomass Bioenergy (ELSEVIER) 56:382–391

Bhattacharyya S, Mitra A, Raha AK (2015) Stored carbon in above ground biomass of dominant mangrove floral species in Sagar island of Indian Sundarbans. J Chem Biol Phys Sci 5(4):4664–4672

Bunting P, Rosenqvist A, Lucas RM, Rebelo LM, Hilarides L, Thomas N, Hardy A, Itoh T, Shimada M, Finlayson CM (2018) The global mangrove watch—a new 2010 global baseline of mangrove extent. Remote Sens 10:1669

Chakraborty A, Datta U, Sufia Z, Pal N, Pramanick P, Mitra A, Mitra A (2016) Spatio-temporal variation of stored carbon in *Sonneratia apetala*: a case study from Indian Sundarbans. Int J Innov Res Technol Sci Eng 5(3):2975–2980

Chastain SG, Kohfeld K, Pellatt MG (2018) Carbon stocks and accumulation rates in salt marshes of the Pacificcoast of Canada. Biogeosci Discuss. https://doi.org/10.5194/bg-2018-166

Chaudhuri AB, Choudhury A (1994) Mangroves of the Sundarbans. The World Conservation Union, Dhaka

Chidumaya EN (1990) Above ground woody biomass structure and productivity in a Zambezian woodland. For Ecol Manag 36:33–46

Connor RF, Gail LC, Beth Beecher C (2001) Carbon sequestration in bay of Fundy salt marshes: implications for restoration of reclaimed marshes. Glob Biogeochem Cycles 1–12

Cooray PLIGM, Kodikara KAS, Kumara MP, Jayasinghe UI, Madarasinghe SK, Dahdouh-Guebas F et al (2021) Climate and intertidal zonation drive variability in the carbon stocks of Sri Lankan mangrove forests. Geoderma 389:114929

Dagar JC, Singh NT, Mongia AD (1993) Characteristics of mangrove soils and vegetation of Bay Islands in India. In: Lieth H., Masoom AAl (eds) Towards the rational use of high salinity tolerant plants, vol I, pp 59–80

Dhar I, Sengupta G, Biswas S, Sinha M, Mitra A (2021) Salinity: A Major environmental factor in sustainability of the Blue Carbon. J Mech Continua Math Sci 16(11):34–42. https://doi.org/10.26782/jmcms.2021.11.00004

Donato DC, Kauffman JB, Murdiyarso D, Kurnianto S, Stidham M, Kanninen M (2011) Mangroves among the most carbon-rich forests in the tropics. Nat Geosci 4:293–297

Drake K, Halifax H, Adamowicz SC, Craft C (2015) Carbon sequestration in tidal salt marshes of the Northeast United States. Environ Manag 56:998–1008

Duarte CM, Losada IJ, Hendriks IE, Mazarrasa I, Marbà N (2013) The role of coastal plant communities for climate change mitigation and adaptation. Nat Clim Chang 3:961–968

Dutta J, Thakur TK, Zaman S, Mitra A (2020) Regulatory effects of salinity on the mangrove biomass: a brief commentary From Indian Sundarbans. eJ App Forest Ecol 8(1):20–27

Ezcurra P, Ezcurra E, Garcillán PP, Costa MT, Aburto-Oropeza O (2016) Coastal landforms and accumulation of mangrove peat increase carbon sequestration and storage. Proc Natl Acad Sci 113:4404–4409

Fourqurean JW, Duarte CM, Kennedy H, Marbà N, Holmer M, Mateo MA et al (2012) Seagrass ecosystems as a globally significant carbon stock. Nat Geosci 5:505–509

http://www.bsienvis.nic.in/Database/IndianMangroves_3941.aspx#WORLDDISTRIBUTIONOFMANGROVES

https://www.fao.org/3/w4095e/w4095e0c.htm

Hu T, Zhang YY, Su Y, Zheng Y, Lin G, Guo Q (2020) Mapping the global mangrove forest aboveground biomass using multisource remote sensing data. Remote Sens 12:1690. https://doi.org/10.3390/rs12101690

Husch B, Miller CI, Beers TW (1982) Forest mensuration, 3rd edn. Wiley, New York. 402

Intergovernmental Panel on Climate Change (IPCC) (2014) AR5 synthesis report: climate change 2014. IPCC, Geneva, Switzerland. https://www.ipcc.ch/report/ar5/syr/

Jagtap TG (1985) Ecological studies in relation to the mangrove environment along the Goa Coast, India, PhD thesis. Shivaji University, Kolhapur, India, 37–44

Jayatissa LP, Dahdouh-Guebas F, Koedam N (2002) A review of the floral composition and distribution of mangroves in Sri Lanka. Bot J Linn Soc 138(1):29–43

Kathiresan K (2008) Biodiversity of mangrove ecosystems. In: Proceedings of mangrove workshop. GEER Foundation, Gujarat, India

Kathiresan K, Rajendran N (2005) Mangrove ecosystems of the Indian Ocean region. Ind J Mar Sci 34:104–113

Kauffman BJ, Arifanti VB, Hernández Trejo H, Jesús García MDC, Norfolk J, Cifuentes M et al (2017) The jumbo carbon footprint of a shrimp: carbon losses from mangrove deforestation. Front Ecol Environ 15:183–188

Mandal RN, Naskar KR (2008) Diversity and classification of Indian mangroves: a review. Trop Ecol 49:131–146

McKee KL, Cahoon DR, Feller IC (2007) Caribbean mangroves adjust to rising sea level through biotic controls on change in soil elevation. Glob Ecol Biogeogr 16:545–556

Mitra A (2018) Salinity: a primary growth driver of mangrove forest. Sustain For 1:1–9

Mitra A (2013) Sensitivity of mangrove ecosystem to changing climate. Springer, New Delhi, Heidelberg, New York, Dordrecht, London. ISBN-10: 8132215087; ISBN-13: 978-8132215080. ISBN 978-81-322-1509-7 (eBook), XIX: pp 323. https://doi.org/10.1007/978-81-322-1509-7

Mitra A (2020) Mangrove forests in India: Exploring ecosystem services. Springer eBook ISBN 978-3-030-20595-9 XV, pp 361. https://doi.org/10.1007/978-3-030-20595-9

Mitra A, Banerjee K (2005) Living resources of the Sea: Focus Indian Sundarbans. WWF-India, Sundarbans Landscape Project, Canning Field Office, West Bengal, India

Mitra A, Banerjee K, Raha AK (2010) Mangroves: a unique sink of carbon in the Tropics; Publicity Division, Directorate of Forests, Govt. of West Bengal 1–32

Mitra A, Nayak B, Pramanick P, Roy M, Chakraborty S, Banerjee K, Amin G, Zaman S (2014) Heavy metal level in a mangrove associate species *Ipomoea pes-caprae* in and around the World Heritage Site of Indian Sundarbans. J Environ Sci Comp Sci Eng Technol 4(1):1–7

Mitra A, Sengupta K, Banerjee K (2009) Pilot analysis of carbon sequestration by dominant mangrove species in Indian Sundarbans. In: Proceedings of the all India seminar on global warming—corporate responsibility. Institute of Engineers (India) on 11 and 12th November 92–102

Mitra A, Sengupta K, Banerjee K (2011) Standing biomass and carbon storage of above-ground structures in dominant mangrove trees in the Sundarbans. For Ecol Manag (ELSEVIER) 261(7):1325–1335

Mitra A, Zaman S, Pramanick P (2022) Blue economy in Indian Sundarbans: exploring livelihood opportunities. Springer eBook ISBN 978-3-031-07908-5 XIV: pp 403. https://doi.org/10.1007/978-3-031-07908-5

Mitra, A, Gati RC (2015) Carbon census in the mangrove ecosystem of lower Gangetic delta. Economol J V:11–27

Naskar KR (2004) Manual of India mangroves. Daya Publishing House, Delhi

Raha AK, Bhattacharyya SB, Zaman S, Banerjee K, Sengupta K, Sinha S, Sett S, Chakraborty S, Datta S, Dasgupta S, Roy Chowdhury M, Ghosh R, Mondal K, Pramanick P, Mitra A (2013) Carbon census in dominant mangroves of Indian Sundarbans. The J Energy Environ Sci 127:345–354

Sanjappa M, Venu P, Arisdasan W, Albertson WD (2011) A review of mangrove species in India. In: Bhatt JR, Macintosh Dj, Nayar TS, Pandey CN, Nilaratna BP (eds) Towards conservation and management of mangrove ecosystems in India. IUCN India, pp 35–50

Selvam V, Gnanappazham L, Navamuniyammal M, Ravichandran K, Karunagaran VM (2002) In: Atlas of mangrove Wetlands of India Part 1-Tamil Nadu. M.S. Swaminathan Research Foundation, Taramani, Chennai, pp 58

Sengupta K, Roy Chowdhury M, Bhattacharyya SB, Raha AK, Zaman S, Mitra A (2013) Spatial variation of stored carbon in *Avicennia alba* of Indian Sundarbans. Discov Nat 3(8):19–24

Singh VP, Garge A (1993) In: Ecology of mangrove swamps of the Andaman Islands. International Book Distributors (Dehradun, India), pp 181

Trivedi S, Zaman S, Ray Chaudhuri T, Pramanick P, Fazli P, Amin G, Mitra A (2016) Inter-annual variation of salinity in Indian Sundarbans. Ind J Geo-Mar Sci 45(3):410–415

Twilley RR, Chen RH, Hargis T (1992) Carbon sinks in mangroves and their implications to carbon budget of tropical coastal ecosystems. Water Air Soil Pollut 64:265–288

Zaman S, Bhattacharyya SB, Pramanick P, Raha AK, Chakraborty S, Mitra A (2013) Rising water salinity: a threat to mangroves of Indian Sundarbans. In: Abedin A, Habiba U, Shaw R (eds) Water insecurity: a social dilemma (Edited by) published by Emerald Group Publishing Limited (ISBN: 9781781908822)

Zeng Y, Friess DA, Sasrira TV, Siman K, Koh LP (2021) Global potential and limits of mangrove blue carbon for climate change mitigation. Curr Bio 31:1737–1743

The manufacturer's authorised representative in the EU is Springer Nature Customer Service Centre GmbH, Europaplatz 3, 69115 Heidelberg, Germany. If you have any concerns regarding our products, please contact ProductSafety@springernature.com

Printed and bound by CPI Group (UK) Ltd, Croydon, CR0 4YY
01/10/2026
02225855-0008